A Technical Guide For Performing and Writing Phase I Environmental Site Assessments

A Technical Guide For Performing and Writing Phase I Environmental Site Assessments

Thomas M. Socha, M.S.

Writers Club Press
San Jose New York Lincoln Shanghai

A Technical Guide For Performing and Writing Phase I Environmental Site Assessments

All Rights Reserved © 2001 by Thomas M. Socha

No part of this book may be reproduced or transmitted in any form or by any means, graphic, electronic, or mechanical, including photocopying, recording, taping, or by any information storage retrieval system, without the permission in writing from the publisher.

Writers Club Press
an imprint of iUniverse.com, Inc.

For information address:
iUniverse.com, Inc.
5220 S 16th, Ste. 200
Lincoln, NE 68512
www.iuniverse.com

The reader should not rely on this publication to address specific questions that apply to a particular set of facts. The author and publisher make no representation or warranty, express or implied, as to the completeness, correctness, or utility of the information in this publication. In addition, the author and publisher assume no liability of any kind whatsoever resulting from the use of or reliance upon the contents of this book.

ISBN: 0-595-19929-1

Printed in the United States of America

In memory of my Grandparents.

Contents

Acknowledgements xi
List of Abbreviations xiii
Introduction xv
Chapter 1: Know Yourself, Your Staff, and Your Client 1
    INTRODUCTION 1
        Know Yourself As A Manager 1
        Know Your Staff 2
        Know Your Client 3
Chapter 2: Management of an Environmental Site Assessment Step-by-Step .. 4
    INTRODUCTION 4
        Project Management 6
        Contract 7
    PRE-PHASE I ENVIRONMENTAL SITE ASSESSMENT 15
        Scheduling 15
        Background Information 15
        Health and Safety Plan 16
    SITE RECONNAISSANCE 16
        Opening Conference or Meeting 17
        Physical and Visual Inspections 17
        Site-Visit Interviews 19
        Site-Visit Records Review 23
        Closing Conference or Meeting 24
        Team Meeting(s) 24
    POST SITE-VISIT DATA GATHERING ACTIVITIES 24

Documentation 24
STATE, FEDERAL, AND LOCAL RECORD REVIEWS 25
Required Environmental Information 25
Physical or Environmental Setting 28
Local Government Interviews 28
Historical Record Review 28
PHASE I ENVIRONMENTAL SITE ASSESSMENT 29
Reporting Phase I ESA Findings 29
Finalizing Audit Findings 29
Draft Assessment Report 31
Final Phase I Environmental Site Assessment Report 31
Chapter 3: From Coverletter to Executive Summary 32
INTRODUCTION 32
Coverletter 32
Title Page 33
Report Introduction 34
EXECUTIVE SUMMARY 35
Brief Description of the Subject property 36
Historical Review Summary 36
Site Reconnaissance 37
Regulatory Review 37
Data Failure 37
Findings and Conclusions 38
Table Of Contents 41
Limitations 44
Preparation 44
Additions or Deletions 45
Warranties, Guarantees, and Certification 45

Chapter 4: From Historical Record Reviews to Map of the Subject Property ..50
    HISTORICAL RECORDS REVIEW 50
        Chain of Title Search 52
        Information Regarding Environmental Liens 52
        Aerial Photographs 52
        Fire Insurance Maps 54
        City Directories 56
        Interviews 58
        Interviews with Knowledgeable Site Contacts 59
        Property Tax Records 59
        Previous Site Investigations 60
        Other Sources 62
    SITE RECONNAISSANCE 62
        Name, Date, and Time of Reconnaissance 63
        Weather 63
        Visual and Physical Observations 63
        Current Property Use 64
        Intended Property Use 64
        Utilities 65
        Water Wells 65
        Sanitary Sewer/Septic System 66
    BUILDING INTERIOR 66
        Interior Waste Streams 68
    BUILDING EXTERIOR 68
        Exterior Site Operations 69
        Hazardous Waste Disposal 69
        Agricultural Waste 70
        Solid Waste Stream and Disposal 70
        Underground and Above Ground Storage Tanks 70

x • *A Technical Guide For Performing and Writing Phase I Environmental Site Assessments*

    Pits, Ponds, Lagoons and Waste Disposal Areas 71
    Polychlorinated Biphenyls 71
    Electromagnetic Fields 73
    Vegetation 73
    Oil and Gas Wells 74
    Topography 74
    Drainage Patterns 75
    General Soil Profile 75
    Geology 76
    Local Groundwater Flow 76
    Legal Description of Property 77
    ADJOINING PROPERTIES 77
    REGULATORY FILE REVIEW 78
    Local Building Department 89
    Local Zoning Department 89
    Map 89

Chapter 5: From Non-Scope Sources to Documentation of Sources 91
    NON-SCOPE AND ADDITIONAL SOURCES 91
    Asbestos Containing Materials Inspection 91
    MULTI-MEDIA COMPLIANCE INSPECTION 95
    CONCLUSIONS AND RECOMMENDATIONS 97
    Documentation of Sources 100

Appendix A: Chain of Custody 103

Appendix B: Environmental Assessment Checklist 105

Appendix C: References 109

Bibliography 111

Index 113

About the Author 115

ACKNOWLEDGEMENTS

I would like to acknowledge the following people:

My parents Gerald and Josephine Socha.

My brothers Paul and Bob Socha.

My friends Tom and Bonnie Kapp, Dale and Bethany Coats, Gary Coats, and Joe Kapusnak.

I would like to acknowledge the following institutions:

Wayne State University for having extensive research libraries.

Oakland Community College for having a great selection of technical programs.

And, I like to thank all the people who believed in my abilities and skills. However, I would especially like to thank the people, corporations, and institutions that did not believe because without them I would not have written this book.

LIST OF ABBREVIATIONS

ASTM—American Society for Testing and Materials.
ACM—Asbestos Containing Building Material
AHERA—Asbestos Hazard Emergency Response Act of 1986.
CAA—Clean Air Act.
ERCLA—Comprehensive Environmental Response, Compensation and Liability Act of 1980 (as amended, 42 USC § 9601 et seq.).
CERCLIS—Comprehensive Environmental Response, Compensation, and Liability Information System.
CORRACTS-TSD—Facilities subject to Corrective Actions under RCRA.
EPA—United States Environmental Protection Agency
ERNS—Emergency Response Notification System
FOIA—Freedom of Information Act.
LUST—Leaking Underground Storage Tank List
MSDS—Material Safety Data Sheet.
NESHAPs—National Emission Standards for Hazardous Air Pollutants (CAA)
NPDES—National Pollutant Discharge Elimination System.
NPL—National Priorities List
PCB—Polychlorinated biphenyl
PPM—Parts per million
RACM—Regulated Asbestos Containing Materials.
RCRA—Resource Conservation and Recovery Act (as amended, 42 USC § 6901 et seq.).
RCRA Generators—RCRA Information System Small and Large Quality Generators
RCRIS—RCRA Information System Small and Large Quality Generators

RCRIS (TSD)—RCRA Information System Treatment Storage and Disposal Facilities.

SARA—Superfund Amendments and Reauthorization Act of 1986 (amendment to CERCLA).

SPCC—Spill, Prevention, Containment & Counter-measures

SHWS—State Hazardous Waste Sites

SWF/LF—Michigan Solid Waste Facilities

TSCA—Toxic Substance Control Act (15 USC Section 2601 et seq.).

TSDF—Treatment, Storage, Disposal Facility

USC—United States Codes

USGS—Unites States Geological Survey.

UST—State Registered Underground Storage Tanks

INTRODUCTION

With the passage of the 1986 Superfund Amendments and Reauthorization Act (SARA) which amended the Comprehensive Environmental Response, Compensation and Liability Act (CERCLA or Superfund) introduced the "innocent landowners" defense, which required that a purchaser conduct "all appropriate inquiry into the prior history and uses of the property consistent with good commercial or customary practice in an effort to minimize liability" (42 U.S.C. Sec. 9601(35)(B)).

In the late 1980s, after several United States House bills failed to pass a definition for "appropriate inquiry." A small group of lending and real estate organizations met to address this missing definition. This group had chosen the American Society for Testing and Materials (ASTM) to prepare a voluntary consensus standard guidance document. In October 1992, the ASTM's Subcommittee on Commercial Real Estate voted to send its environmental site assessment standard to the 30,000 ASTM members for approval. In May 1993, the E 1528–94 Transaction Screen and E 1527–93 Phase I Site Assessment standards were adopted by ASTM.

Chapter 1: Know Yourself, Your Staff, and Your Client

INTRODUCTION

As a manager (i.e. senior management) you are responsible to know your staff and clients. Many times management is too occupied with the bottom-line, and may desensitize themselves to their staff and client needs. The following sections discuss information that will be useful in determining your staff members and clients needs.

Know Yourself As A Manager

- *Do not flaunt your experience or education.* "I am the Professional Engineer and this is how the project will be done or it is the highway." A better approach is to explain why and how this is done.
- *Do not live in an ivory tower.* Analyze and accept new ideas. Make yourself accessible and approachable to your staff members and be open-minded to new ideas from your staff.
- *Do onto others, as you would want done onto you.* Treat people with respect, and professionalism (no swearing, no hanging-up a phone in a disturbing manner, no misspelling of the staff members names incorrectly on memos).
- *Do not strictly rely on middle management's evaluations of staff members.* Often times the middle management (project engineers, project managers, etc.) might also be intimidated by your leadership behavior and tell you what you expect to hear. They might also have a hidden agenda against an individual. Thus, discuss findings of concern directly with the staff member who is being evaluated or disciplined.

- *Train your staff members.* Remember they may not have the background, skills, and education as you. Training will help improve individual self-worth, as well as, their work and the company's performance. Do not train your staff in a few hours and expect them to have the knowledge you have required. Training new personnel requires leadership, patience, understanding, and time.
- *Do use your experience and education to improve your staff performance.* This includes training new personnel.
- *Encourage your staff members.* If you are going to criticize your employee when they do something wrong. Reward them when the do excellent work. You can catch more flies with honey than with vinegar. Get to know your staff, talk to them as a person and not just as a boss.
- *Do not hang your staff members.* This is accomplished by assigning an entry-level person or staff member a project without proper client information (i.e. time frame, contacts, etc.) from the support staff. Support staff refers to your staff members and management. For example, as a manager if you opt to go on a site walkover with a staff member, you are responsible to help gather and deliver the information to the person who is writing the report.
- *Return and review reports to the staff member in charge of the project in a timely manner.* This is the time to give your input into the report because it is often the manager who will be accompanying the staff member on a site-visit.

Know Your Staff

- Let your staff members know that you care about them as individuals, not only as company employees. Encourage your staff members to succeed and excel.

- Your entry-level and other support staff are neither gods nor mine readers. The entry-level personnel must be trained on company's techniques, styles, etc. from the first day. Tell the staff member in a face-to-face meeting what are their deficiencies and how they can improve their skills.
- A committed company staff member. This is a staff member who will emit to their mistakes without any accuses when confronted. Their job performance will improve with proper training and mentoring. An individual who uses their free time to complete a project (i.e. driving to a site to take some pictures on a Sunday or coming in to the office at 4:30 AM to copy reports) and does not write it down on their time sheet.

Know Your Client

- Be aware of the client's problems and concerns.
- Return all client phone calls and inquiries promptly.
- Be punctual for scheduled meetings and appointments.
- Deliver your services on schedule.
- Maintain regular contact by telephone or in face-to-face conversations.
- Ensure that the client knows that you are working for the client, and that your efforts are focused on the client's need and problems.
- Let the client know that you care about them as a person, not only as a business colleague.

CHAPTER 2: MANAGEMENT OF AN ENVIRONMENTAL SITE ASSESSMENT STEP-BY-STEP

INTRODUCTION

This chapter describes the proper management of the Phase I Environmental Site Assessment (assessment) from pre-contract to submission of the report. The format for the assessment is discussed in this book which is based on the ASTM E 1527-00 (ASTM E-1527), *Standard Practice for Environmental Site Assessments: Phase I Environmental Site Assessment Process*, and ASTM E 1528-00 (ASTM E-1528), *Standard Practice for Environmental Site Assessments: Transaction Screen Process standards*, standard practices.

The ASTM E-1528 and ASTM E-1527 practices are used to "define good commercial and customary practice" in the United States of America, in connection with a parcel of commercial real estate with respect to the range of contaminants within the scope of the CERCLA and petroleum products. These practices are intended to permit a user to satisfy one of the requirements to qualify for the innocent landowner defense to CERCLA liability that is, the practices that constitute "all appropriate inquiry into the previous ownership and uses of the property consistent with good commercial or customary practice" as defined in 42 USC § 9601(35)(B).

Both E-1527 and E-1528 have four main objectives which are:

(1) To unify and put in writing good commercial and customary practice for environmental site assessments for commercial real estate;

(2) To facilitate high quality, standardized environmental site assessments;

(3) To ensure that the standard of appropriate inquiry is practical and reasonable, and;

(4) To clarify an industry standard for appropriate inquiry in an effort to guide legal interpretation of CERCLA's innocent landowner defense.

The ASTM E-1528 practice is intended to address the needs of lenders in rural areas with no previous development and where large areas of undeveloped land exist. ASTM E-1528 transaction screen is based on a questionnaire with 23 questions and is used in the interviewing inquiry of the owner/occupant(s), site-visit (visually and physically), a review of governmental records and historical sources. If any of the questions are answered "affirmative" or "unknown" on the questionnaire, the user must exercise business judgment to determine if an assessment is necessary. The transaction screen can be used without retaining an environmental professional.

An environmental professional must conduct the ASTM E-1527 site assessment. Many of the 23 questions that are asked in ASTM E-1528 transaction screen are examined again but with a more detail search of the property's historical and current uses.

No environmental site assessment can wholly eliminate uncertainty regarding the potential for recognized environmental conditions in connection with a property. By performing the report in accordance of ASTM standard practice ASTM E-1527 its intention is to reduce, but not eliminate, uncertainty regarding the existence of recognized environmental conditions in connection with the subject property, and recognizing reasonable limits of time and money. The ASTM E-1527 requires a review of reasonably ascertainable of standard historical sources, interviews, and records search for the subject property, its surrounding area, adjoining properties, and governmental agencies.

Reasonably ascertainable is defined by ASTM E-1527 as "information that is publicly available, obtainable from source with reasonable time and cost constraints, and practically reviewable."

The environmental site assessment report consists of four key components which include the following:

- Records Review (historical and current on-site records)
- Site Reconnaissance

- Interviews with current owners and occupants of the property
- Report

Another component of the report is the physical or environmental setting which includes the mandatory United States Geological Survey (USGS)-Current 7.5 Minute Topographic Map (or equivalent).

The purpose of the record reviews, site reconnaissance, and interviews is to determine if the subject property has any recognized environmental condition(s) (REC).

The ASTM E-1527 standard defines *recognized environmental condition* "as the presence or likely presence of any hazardous substances or petroleum products under conditions that indicate an existing release, a past release, or a material threat of a release of any hazardous substances or petroleum products into structures on the property or into ground, groundwater, or surface water of the property. The term includes hazardous substances or petroleum products even under conditions in compliance with the laws.

The term is not intended to include de minimis conditions that generally do not present a material risk of harm to public health or the environment and that generally would not be subject of an enforcement action if brought to the attention of appropriate governmental agencies."

Project Management

Before the contract is sign and delivered by the client, it is imperative for the company's leadership to inform the staff members of an appending project.

You just received a signed contract from your client. What are your next steps? You must first fill-out a Chain of Customer Custody (CCC) Sheet. A CCC contains staff personnel, client, and financial institutional (if any) information (Refer to Appendix A). When management receives the contract it is their responsibility to fill-out the client's name and/or financial institution information on the CCC before sending it to the project custodian.

The project custodian (custodian) is the individual who is assigned an assessment from the project manager. The custodian is in charge of the assessment from the walkover to submission of the final document to the client. If the project custodian cannot finish the report then he or she must fill-out a CCC in order to transfer over the project to the new project custodian.

The custodian and/or assessment team is also referred to as the *environmental professional(s)*.

Contract

A contract is an agreement between a company performing a service and its client (Refer to Example 2-1). A general contract contains the following information but, not limit to: the Scope of Services; Responsibilities of the Client; Changes, Delays, Excused Performance; Compensation; Insurance, Dispute Resolution, Allocation of Risk, and Miscellaneous Provisions.

Example 2-1:

TM Socha Company
General conditions for Environmental Audits and Site Assessments

These General Conditions are a part of each agreement between a TM Socha Company and its client for the performance of environmental audits and site assessments. In these General Conditions, the TM Socha Company is called "TMS", the party for whom the services are performed is called "Client", and the written agreement between the parties, including these General Conditions, is called "this Agreement".

Section 1: Services by TMS

1.1 **Scope of services; standard of care.** TMS will perform the services described in this Agreement and in any work release documents or change orders which are issued under this Agreement and signed by both parties. In the course of performing the services, TMS may recommend additional investigations, sampling and/or testing to increase the information upon which its conclusions will be based, and shall be entitled to additional compensation and time for any additional services authorized by Client. Unless expressly agreed in writing, TMS shall not be obligated to remove from the site or to dispose of any petroleum or petroleum products (collectively called "Oil") or any hazardous, toxic, radioactive or infectious substances, including any substances regulated under RCRA or any other federal or state environmental laws (collectively called "Hazardous Materials"). In performing the services TMS will exercise the degree of care and skill ordinarily exercised by reputable companies performing the same or similar services in the same geographic area. TMS will not have any obligation to perform services not expressly described in this Agreement or in work release documents or change orders signed by TMS.

1.2 Report. Any report issued by TMS will set forth its findings and conclusions based on the limited information available from the observations, investigations, sampling and/or testing conducted under this Agreement. In preparing its report, TMS may review and interpret information provided by Client and third parties and shall be entitled to rely on the accuracy of such information, without

Proposal/Agreement NO.:
Name Of Client:
Project Name:

performing an independent verification. TMS shall also be entitled to rely on the accuracy of laboratory results. TMS may include in its report a Statement of Limitations describing the limitations of its investigations and findings and indicating that the report is for Client's use only and shall not be relied upon by any third party, except as expressly agreed in writing by TMS, and then at such third party's own risk.

1.3 No warranty. NO WARRANTIES OR GUARANTIES, EXPRESS OR IMPLIED, ARE MADE WITH RESPECT TO ANY GOODS OR SERVICES PROVIDED UNDER THIS AGREEMENT, AND ANY IMPLIED WARRANTIES OF MERCHANT ABILITY OR FITNESS FOR A PARTICULAR PURPOSE ARE EXPRESSLY DISCLAIMED. Geologic, soil and water conditions are variable and indeterminate in nature and are subject to change over time as a result of natural and man-made processes. Legal requirements and industry and professional standards are also variable and subject to change. In addition, the number of investigations and observations TMS makes, the number of samples it collects, and the number of tests it performs are necessarily limited by budgetary and time constraints, and observations and samples by their very nature are not entirely representative of what is being observed or sampled. TMS therefore does not guaranty that all violations, problems or sources of possible contamination will be identified, that all contaminants will be detected or properly identified, or that requirements, standards or conditions will not change over time.

1.4 Restoration. TMS will exercise reasonable care to minimize damage to the site. However, Client acknowledges that some damage may occur in the normal course of performing the services, even if due care is exercised, and agrees that TMS will not be liable for such damage and will be entitled to additional compensation if it is asked to perform restoration services not expressly included in the scope of services.

1.5 Estimates. Any opinions of probable construction or implementation costs, financial evaluation, feasibility studies or economic analyses prepared by TMS will represent its best judgment based on its experience and available information. However, Client recognizes that TMS has no control over costs of labor, materials, equipment or services furnished by others or over market conditions or contractors' methods of determining prices, and that any evaluation of a facility to be constructed or work to performed is speculative. Accordingly, TMS does not guaranty that proposals, bids or actual costs will not vary from opinions, evaluations or studies submitted by TMS.

1.6 Other contractors. TMS shall not have any duty or authority to direct, supervise or oversee any contractors of Client or their work or to provide the means, methods or sequence of their work or to stop their work. TMS's services and/or presence at a site shall not relieve others of their responsibility to Client or to others. TMS shall not be liable for the failure of Client's contractors or others to fulfill their responsibilities, and Client agrees to indemnify, hold harmless and defend TMS against any claims arising out of such failures.

1.7 Health and safety. TMS shall not be responsible for health or safety programs or precautions related to Client's activities or operations, Clients other contractors, the work of any other person or entity, or Client's site conditions. TMS shall not be responsible for inspecting, observing, reporting or correcting health or safety conditions or deficiencies of Client or others at Client's site. For separate consideration of $10 and other good and valuable consideration, the receipt and adequacy of which are hereby acknowledged, Client agrees to indemnify, hold harmless and defend TMS to the fullest extent permitted by law against any and all claims resulting from or related to bodily injury or death arising out of such conditions or deficiencies or the actions or failure to act of others, regardless of whether TMS is claimed or deemed to have been negligent in connection therewith. So as not to discourage TMS from voluntarily addressing health or safety issues while at Client's site, in the event TMS does address such issues by making observations, reports, suggestions or otherwise, TMS shall nevertheless have no liability or responsibility arising on account thereof, and Client's indemnity set forth above shall apply to any claims arising therefrom.

1.8 Litigation support. TMS will not be obligated to provide expert witness or other litigation support services related to the services, unless expressly agreed in writing. In the event TMS is required to respond to a subpoena, government inquiry or other legal process related to the services in connection with a proceeding to which it is not a party, Client shall reimburse TMS for its costs and compensate TMS at its then standard rates for the time it incurs in gathering information and documents and attending depositions, hearings, and the like.

1.9 Confidential information. Although TMS generally will not disclose without Client's consent information received by Client or developed during the course of its services and designated by Client as confidential (but not including information which is publicly available, is already in TMS possession, or is obtained from third parties), TMS shall not be liable for disclosing such information if it in good faith believes such disclosure is required by law or is necessary to protect the safety, health, property or welfare of the public. TMS shall notify Client (in advance, except in emergency) of such disclosure.

Section 2: Responsibilities of Client

2.1 Client Requirements. Client, without cost to TMS, shall:
(a) Designate to TMS in writing a person to act as Client's representative with respect to the services
(b) Provide or arrange for access and make all provisions for TMS to enter each site and the buildings thereon at reasonable times throughout contract performance
(c) Comply with all laws and provide any notices required to be given to any government authorities in connection with the services, except for such notices TMS has expressly agreed in writing to give.
(d) Before commencement of any drilling or excavation at a site, furnish TMS with a complete description (including the accurate location) of all underground objects and structures at the site, including but not limited to wells, tanks and utilities; and indemnify, hold harmless and defend TMS against claims arising out of damages to underground objects or structures not properly identified
(e) Furnish TMS with all approvals, permits and consents from government authorities and others as may be required for performance of the services, except for those TMS has expressly agreed in writing to obtain
(f) Inform the owner of the site (if different from Client) of any contamination by or release of Oil or Hazardous Materials at the site.

2.2 Information. Before commencement of any services at a site, Client, without cost to TMS, shall provide TMS with the following items to the extent they are reasonably available to Client, except for any items TMS has expressly agreed in writing to obtain:
(a) Any information pertinent to the services
(b) Information concerning prior owners of the site and any current or historical uses of or activities on the site by Client, prior owners or others
(c) A legal description of the site and all available surveys and site plans and other information concerning site conditions, topography, easements, rights of way, and zoning, deed or land use restrictions
(d) Material Safety Data Sheets (MSDSs) that conform to OSHA requirements related to all Hazardous Materials located at the site
(e) Contingency plans related to the site

(f) Previous environmental audits and/or assessments related to the site
(g) Environmental permits related to the site
(h) All other information concerning known or suspected Hazardous Materials at the site, contamination of the site by Oil or Hazardous Materials, including the nature of the contaminant and the quantity, location and date of any release, and any conditions at the site requiring special care.

2.3 Manifests. Before TMS removes from a site any Oil or Hazardous Materials, Client shall sign any required hazardous waste manifests in conformance with all DOT and other government regulations, listing, Client as generator of the waste. If someone other than Client is the generator of the waste, Client shall arrange for such other person to sign such manifests. TMS shall not directly or indirectly assume title to or own any materials handled or removed from any site, including Oil or Hazardous Materials. Nothing in this Agreement shall be construed to make TMS a "generator" as defined in RCRA or any similar laws governing the treatment, storage or disposal of waste. Except to the extent TMS's responsibilities expressly include identification of the waste, Client shall provide waste material profiles that accurately characterize the waste.

2.4 Confidentiality. Client acknowledges that the technical and pricing information contained in this Agreement is confidential and proprietary to TMS and agrees not to disclose it or otherwise make it available to others without TMS's express written consent.

2.5 Documents. All reports, notes, calculations, data drawings, estimates, specifications and other documents and computerized materials prepared by TMS are instruments of TMS's services and shall remain TMS property. Documents or computerized materials provided to Client are for Client's use only for the purposes disclosed to TMS, and Client shall not transfer them to others or use them for any purpose for which they were not prepared, without TMS's prior express written consent; provided, however, that TMS will not unreasonably withhold its consent for Client to use TMS's report or provide it to others in furtherance of Client's remediation of the site. Any use of or reliance on TMS's report by a third party, even with TMS's consent, shall be at such party's own risk.

Section 3: Changes; Delays; Excused Performance

3.1 Changes. Unless this Agreement expressly provides otherwise, TMS's proposed compensation represents its best estimate, taking into account the costs, effort and time it expects to expand in performing the services as it currently understands them to be, based on reasonable assumption of the conditions and circumstances under which the services will be performed. As the services are performed, conditions may change or circumstances outside TMS's reasonable control (including changes in law) may develop which would require TMS to expend additional costs, effort or time to complete the services, in which case TMS will notify Client and an equitable adjustment will be made to TMS's proposed compensation is based on the assumption that TMS will not encounter and underground structures, utilities, boulders, rock, water, running sand or other unanticipated conditions in the course of drilling or excavation, and TMS shall be compensated for any additional efforts expended or costs incurred in addressing such conditions. If unusually hazardous conditions not reasonably anticipated are discovered at the site, TMS in its sole discretion may suspend or terminate the services. In the event such conditions or any other conditions or circumstances justify the suspension or termination of services, TMS shall be compensated for services previously performed and for costs reasonably incurred in connection with the suspension or termination.

3.2 Force majeure. TMS shall not be responsible for any delay or failure of performance caused by fire or other casualty, labor dispute, government or military action, transportation delay, inclement weather, Act of God, act or omission of Client or its contractors, failure of Client or any government authority timely to review or to approve the services or to grant permits or approvals, or any other cause beyond TMS's reasonable control, and TMS's compensation shall be equitably adjusted to compensate it for any additional costs it incurs due to any such delay.

Section 4: Compensation

4.1 Rates. Unless otherwise agreed in writing, TMS shall be compensated for its services at its standard rates for professional services, labor, equipment and materials, and shall be reimbursed for its out-of-pocket expenses (plus reasonable profit and over-head).

4.2 Invoices. TMS may invoice Client on a monthly or other progress-billing basis. Invoices are due and payable upon receipt by Client. On amounts not paid within 30 days of invoice date, Client shall pay interest of 1.5% per month or, if less, the maximum rate allowed by law. If Client disagrees with any portion of an invoice, it shall notify TMS in writing of the amount dispute and the reason for its disagreement within 21 days of receipt of the invoice, and shall pay the portion not in dispute.

4.3 Suspension, etc. If payment is not received within 45 days of the invoice date, TMS may upon 7 days' notice suspend or terminate the services and receive compensation for services previously performed and for costs reasonably incurred in connection with the suspension or termination.

4.4 Collection. Client shall reimburse TMS for its costs and expenses (including reasonable attorneys' and witnesses' fees) incurred in any litigation for collection under this Agreement in which TMS obtains a judgment in its favor.

4.5 Taxes, etc. Unless expressly agreed in writing, TMS's fees do not include any taxes, excises, fees, duties or other government charges related to the goods or services provided under this Agreement, and Client shall pay such amounts or reimburse TMS for any amounts it pays. If Client claims that any goods or services are subject to a tax exemption, it shall provide TMS with a valid exemption certificate.

Section 5: Insurance; Dispute Resolution; Allocation of Risk

5.1 Insurance. TMS will maintain workers compensation insurance as required by law; employers liability, comprehensive general liability and automobile liability insurance each with coverage of $1 million per occurrence; and professional liability insurance with coverage of $1 per claim; and upon request will furnish insurance certificates to Client. TMS will purchase additional insurance if requested by Client, provided the insurance is reasonably available from carriers acceptable to TMS and Client reimburses TMS for its costs.

5.2 Disputes. If a claim or dispute arises out of this Agreement or its performance, the parties agree to endeavor in good faith to resolve it equitably through negotiation or, if that fails, through nonbinding mediation under the rules of the American Arbitration Association, before having recourse to the courts. However, prior to or during negotiation or mediation, either party may initiate litigation that would otherwise become barred by a statute of limitations, and TMS may pursue any property liens or other rights it may have to obtain security for the payment of its invoices.

5.3 Indemnification. Client agrees to indemnify, hold harmless and defend TMS from and against any and all liabilities, demands, claims, fines, penalties, damages, forfeitures and suits, together with reasonable attorneys' and witnesses' fees and other costs and expenses of defense and settlement, which TMS may incur, become responsible for or pay out as a result of death or bodily injury or threat thereof to any person, destruction or damage to any property, contamination of or adverse effect on natural resources or the environment, any violation of local, state or federal laws, regulations or orders, or any other damages claimed by third parties (collectively, "Damages") based on or arising in whole or in part out of TMS's performance under this agreement or out of Client's violation of law breach of this Agreement; provided, however, that Client shall not be obligated to indemnify TMS to the extent such Damages are caused directly by the negligence or willful misconduct of TMS.

5.4 Indemnification regarding hazardous materials. Client acknowledges that TMS does not have any responsibility for preexisting Oil or Hazardous Materials at the site or (expect to the extent necessary to take and analyze samples) for their monitoring, handling, storage, transportation, disposal or treatment, that TMS's compensation is not commensurate with the unusually high risks associated with such materials, and that insurance is not reasonably available to protect against such risks. Therefore, for separate consideration of $10 and other good and valuable consideration, the receipt and adequacy of which are hereby acknowledge, and in addition to the indemnification provided in Section 5.3, Client agrees to indemnify, hold harmless and defend TMS against all Damages such as costs of response or remediation arising out of the application of common law or statues such as CERCLA or other "Superfund" laws imposing strict liability or Damage arising out of TMS's negligence; provided, however, that Client shall not be required to indemnify, hold harmless or defend TMS to the extent such Damages are caused directly by TMS's gross negligence or willful misconduct.

5.5 Limitation of liability. TMS's liability for any and all claims arising out of this Agreement or out of any goods or services furnished under this Agreement, whether based in contract, negligence, strict liability, agency, warranty, trespass, or any other theory of liability, shall be limited to $100,000 or the total compensation received by TMS from Client under this Agreement, whichever is greater. In no event shall TMS be liable for special, indirect, incidental or consequential damages, including commercial loss, loss of use, or lost profits, even if TMS has been advised of the possibility of such damages.

5.6 Employee injury. Client agrees not to implead or to bring as action against TMS based on any claim of personal injury or death occurring in the course or scope of the injured or deceased person's employment with TMS and related to the services performed under this Agreement.

5.7 Defense. Any defense of TMS required to be provided by Client under this Agreement shall be with counsel selected by TMS and reasonably acceptable to Client.

Section 6: Miscellaneous Provisions

6.1 Notices. Notices between the parties shall be in writing and shall be hand delivered or sent by certified mail or acknowledge telefax.

6.2 Assignment, etc. Neither Client nor TMS shall assign or transfer any rights or obligations under this Agreement, except that TMS may assign this Agreement to its affiliates and may use subcontractors in the performance of its services. Nothing contained in this Agreement shall be construed to give any rights or benefits to anyone other than Client and TMS, without the express written consent of both parties. The relationship between Client and TMS is that of independent contracting parties, and nothing in this Agreement or the parties' conduct shall be construed to create a relationship of agency, partnership or joint venture.

6.3 Governing law. This Agreement shall be governed by and constructed in accordance with the laws of The State of Michigan.

6.4 Heading. The headings in this Agreement are for convenience only and are not a part of the agreement between the parties.

6.5 Entire agreement, etc. The written document of which these General Conditions are a part is the entire agreement between the parties, and supersedes all prior agreements. Any amendments to this Agreement shall be in writing and signed by both parties. In no event will the printed terms on any purchase order, work order or other document provided by Client modify or amend this Agreement, even if it is signed by TMS, unless TMS signs a written statement expressly indicating that such terms supersede the terms of this Agreement. In the event of an inconsistency between these General Conditions and any other writing that comprise this Agreement, the other writing shall take precedence.

PRE-PHASE I ENVIRONMENTAL SITE ASSESSMENT

This is the process of scheduling the site-visit, designing the proper health and safety plan, if necessary, and gathering the background information such as historical and on-site records. Checklists on what information should be gathered and can be located in Appendix B.

Remember, the custodian is limited by time to complete the report it is usually 20 days from the site-visit to the submission of the report.

The assessment team should only contain one custodian and if possible, at least one site observer.

Scheduling

Assessment activities shall be developed and documented by the custodian. The custodian is responsible for scheduling a pre-visit site orientation meeting, identifying and introducing site contacts, scheduling dates of the site-visit, draft and review report, and issue the final report to the client.

Background Information

Background information shall be used to develop an assessment report or refine an existing assessment report. Prior to the site-visit, the subject property owner, key site manager and user shall be asked if they know whether any of the documents (background information) listed in the next paragraph exists.

Background information may consist of records, process and site descriptions, operation and maintenance manuals, emergency plans, environmental manuals, compliance inspection reports, previous audit reports, environmental site assessment reports, environmental permits, environmental citations, notices of violations and other relevant information.

A pre-assessment questionnaire should be delivered to the key site managers, users, nonresidential occupants, property owner(s), and government officials prior to the site-visit to gather information on current and prior uses of the subject and adjoining properties.

Now is also a good time to send out the "Freedom of Information Act" (FOIA) requests to your federal, state and local governments regarding the subject property, adjoining properties, and surrounding area.

Health and Safety Plan

If needed, a health and safety plan for the site-visit shall be prepared. Any data created by monitoring instruments carried by the assessment team to detect health and safety threats shall not be assessment, unless specified in the assessment plan.

SITE RECONNAISSANCE

The purpose of the site reconnaissance or site-visit is to obtain information indicating the likelihood of identifying recognized environmental conditions in connection with the subject property and from the adjoining properties and surrounding area. Document your findings with a photograph of the potential recognized environmental conditions on the subject and adjoining properties.

The custodian must gather information concerning the current and past uses of the subject property, adjoining properties, and surrounding area, this is accomplished by visual or physical observations, interviews, and record reviews.

Opening Conference or Meeting

An opening conference or meeting should be held, bringing together the assessment team and the subject property entity (i.e. user, property owner, and/or key site manager). A user is the party seeking to use an environmental site assessment of the property. A user may include, without limitation, a purchaser of property, a potential tenant of property, an owner of property, a lender, or a property manager.

ASTM E-1527 defines "a key site manager as a person with good knowledge of the uses and physical characteristics of the property. This person is usually the property manager, the chief physical plant supervisor, or head maintenance person."

The meeting should facilitate the gathering of information by the assessment team and encourage discussion of any questions or concerns. The user should provide an overview of the subject property to the assessment team such as filling out a current and prior use questionnaire.

Also, it is the user(s) of the subject property responsibility to gather the following:

- To check title records for environmental liens by means of reasonably ascertainable recorded land title records.
- Specialized knowledge or experience that is material to recognized environmental conditions in connection with the property.
- Reasons for significantly lower purchase price for the subject property.

Physical and Visual Inspections

When conducting the site-visit, the custodian and/or the assessment team shall visually and physically observe the property and any structure(s) located on the property to the extent not obstructed by bodies of water, adjacent buildings, or other obstacles.

Example 2-2, explains what observations must be explored during the site-visit for any evidence of recognized environmental conditions located on the subject property. Any findings shall be described in the report.

Example 2-2:

Interior and Exterior Observations:

- Storage Tanks—Underground and Aboveground
- Quantities of Hazardous Substances and Petroleum Products
- Odors
- Pools or Sumps Containing Liquids
- Drums and Containers of all Sizes
- Electrical or Hydraulic Equipment—Which are known to contain PCB's or likely to contain PCB's

Interior Observations:

- Heating and Cooling Fuel Sources
- Stains or Corrosion on Floors, Walls, or Ceilings
- Locations of Floor Drains and Sumps

Exterior Observations:

- Pits, Ponds, or Lagoons—On the subject property and adjoining properties
- Areas of Stained Soil or Pavement
- Areas of Stressed Vegetation
- Areas that are apparently filled or graded by non-natural causes (or filled by fill of unknown origin)
- Waste Water or other liquid (including storm water) or any discharge into a drain, ditch, or steam on or adjacent to the subject property
- All Wells (including dry wells, irrigation wells, injection well, abandoned wells, or other wells)
- All on-site septic systems or cesspools

Site-Visit Interviews

Interviews should be conducted to obtain information on the subject and adjoining properties that might indicate recognized environmental conditions in connection with the property.

The following questions should be asked to the user prior or during the site-visit.

Questionnaire:

1a. Is the property used for an industrial use?
1b. Is an adjoining property used for an industrial use?

2a. Did you observe evidence or do you have any prior knowledge that the property has been used for an industrial use in the past?

3a. Is the property used as a gasoline station, motor repair facility, commercial printing facility, dry cleaners, photo developing laboratory, junkyard or a landfill, or as waste treatment, storage, disposal, processing, or recycling facility (if applicable, identify which)?

3b. Is any adjoining property used gasoline station, motor repair facility, commercial printing facility, dry cleaners, photo developing laboratory, junkyard or a landfill, or as waste treatment, storage, disposal, processing, or recycling facility (if applicable, identify which)?

4a. Did you observe evidence or do you have any prior knowledge that the property has been used as a gasoline station, motor repair facility, commercial printing facility, dry cleaners, photo developing laboratory, junkyard or a landfill, or as waste treatment, storage, disposal, processing, or recycling facility (if applicable, identify which)?

4b. Did you observe evidence or do you have any prior knowledge that any adjoining property has been used as a gasoline station, motor repair facility, commercial printing facility, dry cleaners, photo developing laboratory, junkyard or a landfill, or as waste treatment, storage, disposal, processing, or recycling facility (if applicable, identify which)?

5a. Are there currently any damage or discarded automotive or industrial batteries, pesticides, paints, or other chemicals in individual containers of >5 gal (19L) in volume or 50 gal (190L) in the aggregate, stored on or used at the property or at the facility?

5b. Did you observe evidence or do you have any prior knowledge that there have been previously any damaged or discarded automotive or industrial batteries, or pesticides, paints, or other chemicals in individual containers of >5 gal (19 L) in volume or 50 gal (190 L) in the aggregate, stored on or use at the property or at the facility

6a. Are there currently any industrial drums (typically 55 gal (208 L) or sacks of chemicals located on the property or at the facility?

6b. Did you observe evidence or do you have any prior knowledge that there been previously any industrial drums (typically 55 gal (208L) or sacks of chemicals located on the property or at the facility?

7a. Did you observe evidence or do you have any prior knowledge that fill dirt has been brought onto the property that originated from a contaminated site?

7b. Did you observe evidence or do you have any prior knowledge that fill dirt has been brought onto the property that is of an unknown origin?

8a. Are there currently any pits, ponds, or lagoons located on the property in connection with waste treatment or waste disposal?

8b. Did you observe evidence or do you have any prior knowledge that there have been previously, any pits, ponds, or lagoons located on the property in connection with waste treatment or waste disposal?

9a. Is there currently any stained soil on the property?

9b. Did you observe evidence or do you have any prior knowledge that there has been previously, any stained soil on the property?

10a. Are there currently any registered or unregistered storage tanks (above or underground) located on the property?

10b. Did you observe evidence or do you have any prior knowledge that there have been previously, any registered or unregistered storage tanks (above or underground) located on the property?

11a. Are there currently any vent pipes, fill pipes, or access ways indicating a fill pipe protruding from the ground on the property or adjacent to any structure located on the property?

11b. Did you observe evidence or do you have any prior knowledge that there have been previously, any vent pipes, fill pipes, or access ways indicating a fill pipe protruding from the ground on the property or adjacent to any structure located on the property?

12a. Are there currently any flooring, drains, or walls located within the facility that are stained by substances other than water or were emitting foul odors?

12b. Did you observe evidence or do you have any prior knowledge that there have been previously any flooring, drains, or walls located within the facility that are stained by substances other than water or were emitting foul odors?

13a. If the property is served by a private well or non-public water system, is there evidence or do you have prior knowledge that contaminants have been identified in the well or system that exceed guidelines applicable to water system?

13b. If the property is served by a private well or non-public water system, is there evidence or do you have prior knowledge that the well has been designed as contaminated by any government environmental/health agency?

14 Does the owner or occupant of the property have any knowledge of environmental liens or governmental notification relating to past or recurrent violations of environmental laws with respect to the property or any facility located on the property?

15a. Has the owner or occupant of the property been informed of the past existence of hazardous substances or petroleum products with respect to the property or any facility located on the property?

15b. Has the owner or occupant of the property been informed of the current existence of hazardous substances or petroleum products with respect to the property or any facility located on the property?

15c. Has the owner or occupant of the property been informed of the past existence of environmental violations with respect to the property or any facility located on the property?

15d. Has the owner or occupant of the property been informed of the current existence of environmental violations with respect to the property or any facility located on the property?

16. Does the owner or occupant of the property have any knowledge of any environmental site assessment of the property or facility that indicated the presence of hazardous substances or petroleum products on, or contamination of the property or recommended further assessment of the property?

17. Does the owner or occupant of the property know of any past, threatened, or pending lawsuits or administrative proceedings concerning a release or threatened release of any hazardous substance or petroleum products involving the property by any owner or occupant of the property?

18a. Does the property discharge wastewater, on or adjacent to property, other than storm water, into a storm water sewer system?

18b. Does the property discharge wastewater, on or adjacent to property, other than storm water, into a sanitary sewer system?

19. Did you observe evidence or do you have any prior knowledge that any hazardous substances or petroleum products, unidentified waste materials, tires, automotive or industrial batteries, or any other waste materials have been dumped above grade, buried and/or burned on the property?

20. Is there a transformer, capacitor, or any hydraulic equipment for which there are any records indicating the presence of PCBs?

21. Do any of the following Federal governmental record systems list the property or any property within the circumference of the area noted below?
National Priorities List-within 1.0 mile (1.6 km)?
CERCLIS List-within 0.5 mile (0.8 km)?
RCRA CORRACTS Facilities-within 1.0 mile (1.6 km)?
RCRA non-CORRACTS TSD Facilities-within 0.5 mile (0.8 km)?

22. Do any of the following state records systems list the property or any property within the circumference of the area noted below:
List maintained by the state environmental agency of hazardous waste sites identified for investigation or remediation that is the state equivalent to NPL-within approximately 1.0 mile (1.6 km)

List maintained by state environmental agency of sites identified for investigation or remediation that is the state equivalent to CERCLIS-within 0.5 mile (0.8 km)?
Leaking Underground Storage Tank (LUST) List-within 0.5 mile (0.8 km)?
Solid Waste/Landfill Facilities-within 0.5 mile (0.8 km)?

23. Based upon review of fire insurance maps or consultation with the local fire department serving the property, all as specified in the guide, are any buildings or other improvements on the property or on adjoining property identified as having been used for an industrial use or uses likely to lead to contamination of the property?

Site-Visit Records Review

The custodian and/or assessment team member should review the on-site records from the following:

- Chemical Inventory
- Material Safety Data Sheets
- Community Right-to-Know Records
- Process Flow Charts
- Transport Manifests
- Safety and Health Audits
- Purchases Records
- Corporate Management Records
- Preparedness and Prevention Plan
- Spill Prevention, Countermeasure, and Control Plan
- Reports Regarding Hydrogeologic Conditions on the Subject Property or Surrounding Area
- Environmental Permits
- Notices or Other Correspondences from any Government Agency
- Hazardous Waste Generator Notices or Reports
- Geotechnical Studies
- Environmental Site Assessment Reports
- Environmental Audit Reports

Closing Conference or Meeting

At the conclusion of the site-visit, a closing conference or meeting should be held to discuss any findings of the recognized environmental conditions to incorporate into the report.

The closing conference provides an opportunity for the assessment team and custodian to better understand, and raise questions about, and draft assessment findings. Post-visit procedures should be discussed at the closing conference including a conflict resolution strategy for assessment findings that are challenged and for closing any open issues.

Also, this is a good time to sketch a map of the area which includes the subject property and adjoining properties.

Team Meeting(s)

Meetings of the assessment team should be conducted to ensure timely and consistent completion of the assessment.

POST SITE-VISIT DATA GATHERING ACTIVITIES

Documentation

Assessment protocols should be completed or explanations provided for open issues (e.g. data failure), in accordance with the assessment using the ATSM E-1527 and ATSM E-1528 standards.

STATE, FEDERAL, AND LOCAL RECORD REVIEWS

The custodian is responsible to gather environmental and historical records on the subject property and surrounding area. Usually the custodian will have a database research company perform an environmental records review.

These database research companies include, but limited to:

Environmental Data Resources, Inc.
3530 Post Road
Southport, CT 06490
Phone: 1-800-352-0050
Fax:1-800-231-6802

VISTA
5060 Shoreham Drive
San Diego, CA 92122
Phone: 1-800-767-0403
Fax: 858-450-6195
Email: info@vistainfo.com
Visit our Web site at http://www.vistainfo.com

Required Environmental Information

The environmental record sources shall include the following with these approximate minimum search distances:
- Federal NLP Site List (1 mile)
- Federal CERCLIS List (0.5 mile)
- Federal RCRA TSD Facilities List (1 mile)
- Federal RCRA generators List (Property and adjoining properties)
- Federal ERNS List (Property only)
- State Lists of Hazardous Waste Sites
- State Landfill and/or Solid Waste Disposal (0.5 mile)
- State Leaking Underground Storage Tanks List (0.5 mile)
- State Registered Underground Storage Tanks Lists (Property and adjoining property)

Example 2-3, describes one governmental record source in connection with the subject property and surrounding areas. Below is a list of local and state record sources that could be checked in addition to the standard environmental sources.

- Lists of Landfills/Solid Waste Disposal Sites
- Lists of Hazardous Waste/Contaminated Sites
- Lists of Registered Underground Storage Tanks
- Records of Emergency Release Reports (SARA § 304)
- Records of Contaminated Public Wells

- Department of Health/Environmental Division
- Fire Department
- Planning Department
- Building Permit/Inspection Department
- Local/Regional Pollution Control Agency
- Local/Regional Water Quality Agency
- Local Electrical Utility Companies (for records relating to PCBs)

Example 2-3:

Michigan Department of Environmental Quality

TMS was only able to search five of the eighteen sites that were listed above according to Ms. B. Smith at the MDEQ. Some documentation from the MDEQ regarding information on the given LUSTs sites are included in Appendix C-3.

TMS's research discovered the following information:

Subject Property

No MDEQ records exist regarding the subject property at the time of the inquiry.

South of the Subject Property

The Former WXXX property on 25 or 77 XXX Street was donated eight years ago to the XXX Symphony Orchestra XXX located at 1111 XXX Avenue. The site had a confirmed release on July 1, 1991. This released was confirmed by presence of product in soil. The release occurred from a 10,000 gallon steel tank which contained gasoline according to the MDEQ records. On March 26, 1996 there was another confirmed release of diesel fuel which lead to soil remediation of 40 cubic yards as of July 3, 1996.

The XXX Service Stations #53XX is located at 5555 XXX Avenue. The site had two 550 gallon steel tanks used to store waste oil and fuel oil and one 8,000-gallon steel and three 6,000-fiberglass tanks used to store gasoline. These tanks were excavated from June 19 through June 21 1989. There was a confirmed release on August 30, 1990 according to the MDEQ records. Approximately 200 cubic yards of soil were excavated. The owner of the property is located at XXX OIL, 111 XXX Lake Road, Suite 160, XXX, Michigan.

West of the Subject Property

XXX Investment Co. is located at 4444 XXX Avenue and is assigned the facility LUST number XXX145. The site is currently delisted under the Michigan Department of the State Police, Fire Marshall division. The confirmation of the release was on October 19, 1992 which lead to soil remediation at the site. Other activities at the site have included the removal of underground storage tanks. The owner of the site is XXX Investment Co., P.O. Box 11111, XXX, Michigan 4XXX.

The XXX Patient Family Services site is located at 1111 XXX Avenue, XXX, MI. The confirmation of the release was on February 14, 1992. The State Fire Marshal denied MUSTFA claim because the tanks were not registered. The release number assigned by the Fire Marshal is XXX. The owner of the property is Holy XXX Church, 1111 XXX Street, XXX, Michigan.

North of the Subject Property

The XXX #05-XXX Service Station also known as the XXX site is located at 1111 XXX Avenue. The site had a confirmed release on September 30, 1991 and another on October 3, 1990.

East of the Subject Property

The XXX V.A. Hospital site is located at 1111 XXX Street, XXX MI 4XXX. Conformation of a release was on February 6, 1992.

Physical or Environmental Setting

The report shall contain and describe the subject property's physical setting by means of a current USGS 7.5 Minute Topographic Map (or equivalent) showing the area on which the subject property is located.

Other sources of physical settings include:
- USGS and/or State Geological Survey-Groundwater Maps
- USGS and/or State Geological Survey-Bedrock Geology Maps
- USGS and/or State Geological Survey-Surficial Geology Maps
- Soil Conservation Service-Soil Maps

Local Government Interviews

Interview(s) shall be performed on one staff member from at least one of following local government agencies:

- Local Fire Department
- Local Health Department
- Local or State Environmental Protection Agency

Historical Record Review

Under the ASTM E-1527 practice the following historical sources may be used:

- Aerial photographs
- Fire insurance maps
- Property tax files
- Land title records (although these cannot be the sole historical source consulted)

- Topographic maps
- City directories
- Building department records or zoning/land use records.

PHASE I ENVIRONMENTAL SITE ASSESSMENT

ASTM E-1527 requires "All obvious uses of the property shall be identified from the present, back to the property's obvious first developed use, or back to 1940, which is earlier. This task requires reviewing only as many of the standard historical sources as are necessary, and that are reasonably ascertainable and likely to be useful."

The report shall discuss the physical setting, record review, interview, and conclude if the subject property contains any recognized environmental conditions.

The custodian should perform the assessment in accordance with the ASTM E-1527 standard and/or any lending institution additional guidelines which are intended to establish an appropriate level of inquiry to determine recognized environmental conditions that exist on the subject property.

Reporting Phase I ESA Findings

The assessment shall be prepared accurately, clearly, and concisely to represent the assessment findings, open issues, and deviations, if any, from the assessment.

Finalizing Audit Findings

The assessment findings and conclusions shall be based upon the recognized environmental conditions on the subject property, surrounding area, and adjoining properties. Example 2-4, describes one method of writing the findings and conclusions in the report.

Example 2-4:

We have performed a Phase I Environmental Site Assessment in conformance with the scope and limitations of ASTM Practice E 1527 of 1111 XXXX Avenue, XXX, Michigan, XXX, the subject property. Any exceptions to, or deletions from, this practice are described in Section 2.2 of this report. This assessment has revealed no evidence of recognized environmental conditions in connection with the property except for the following:

The area that 3711 XXX Avenue is located has a Pewamo-Blount-Metamora soil association. This association is nearly level to gently sloping, very poorly drained somewhat poorly drained soils that have a fine to moderately coarse textured subsoil. Thus, there may be an increase contamination migration from these adjacent properties on to the subject property.

1. Subject Property was built in 1881 and renovated in 1915. The building materials may contain asbestos as indicated from the Sanborn Maps.
2. Adjacent Property. The structures on the properties that surrounded the subject property in the past may have a greater risk than the present day uses of the adjacent properties of contamination migrating to the subject property.
   In 1921, the use of the property on the adjacent parcel to the north of the subject property was the XXX Auto School at 3721—3735 XXX Avenue and contained motor testing, offices, classrooms, ignition school, auto repair and machine shop. The use of the property on the adjacent parcel to the south of the subject property was the Garage and repair shop, and 2nd hand auto sales at 3649—3651 XXX Avenue. There was also an Auto service station on the corner of XXX and XXX Avenue. The use of the property on the adjacent parcel to the east of the subject property was a chemical laboratory/vulcanizing at XXX Avenue, auto repair and used car lot at 12-28 XXX, used car lot at 3750—3756 XXX Avenue, auto sales and repair at 3768 XXX Avenue.
   In 1950, the northern property on the adjacent parcel from the subject property had three stores including a Radio Electronic Television School at 3721-3735 XXX Avenue. The use of the property on the adjacent parcel to the east of the subject property was an auto body and paint shop at 28 XXX. A Printing Company was at 3760 XXX Avenue, and used car sales were at 3780 XXX Avenue.
3. Property within a 0.5 and 1.0 mile radius of the subject property. These include one state hazardous waste site, 7 leaking underground storage tank sites, and 10 underground storage tank sites.

Draft Assessment Report

The custodian and the assessment team should develop a draft assessment report in accordance with the ASTM E-1527 practice for review and comment. Comments on a draft assessment report shall be made in a timely manner and returned to the custodian for the final report.

Final Phase I Environmental Site Assessment Report

The final assessment report shall be issued in accordance of the ASTM E-1527 practice and/or any additions by a lending institution.

The report should contain the following:

- Report Format
- Documentation of Sources
- Record Reviews
- Site Reconnaissance
- Interviews
- Credentials of the environmental professional(s)
- Opinion of the environment professional(s) describing recognized environmental conditions
- Findings and Conclusions
- Deviations
- Signature of the environmental professional(s)

Chapter 3: From Coverletter to Executive Summary

INTRODUCTION

This chapter and the following chapters describe how to write a Phase I Environmental Site using the ASTM E-1527 practice guidelines.

Topics discussed in this chapter include:

- Coverletter
- Title Page
- Report Introduction
- Executive Summary
- Table of Contents

Coverletter

A coverletter is submitted to the client along with the final report; it contains the client and financial lending institution information such as client's name and address. It also includes any additions or deletions from the ASTM standard.

Such as proper lending institution language like the following:

"THE REPORT WAS PREPARED IN ACCORDANCE WITH THE MINIMUM REPORTING REQUIREMENTS AND PROCEDURES FOR ENVIRONMENTAL INVESTIGATIONS FOR LENDING INSTITUTION AND THE LENDING INSTITUTION PHASE I ESA CHECKLIST DATED JANUARY 1, 20XX" and in general accordance with the American Society for Testing and Materials (ASTM) Standard Practice for Environmental Site Assessments: Phase I ESA Process (ASTM Designation: E-1527-00).

Title Page

The Title Page shall have the following (Refer to Example 3-1):
- Company name and project number.
- Title (i.e. PHASE I ENVIRONMENTAL SITE ASSESSMENT)
- Location (Name of subject property and address)
- Prepared For Statement (Client and/or lending institution)
- Prepared By Statement
- Project Title that is in italics, capital letters, and bold.
- Date of the Report

Example 3-1:

> PEOPLE TECHNOLOGY, INC., PROJECT NUMBER 20-100
>
> PHASE I ENVIRONMENTAL SITE ASSESSMENT
>
> *Location:*
>
> *Smith Industries*
> *1111 Anywhere Avenue*
> *Somewhere, Michigan 48212*
>
> *Prepared For:*
>
> *Mr. Joe Smith*
> *000 Somewhere Street*
> *Oakland, MI 48071*
>
> *Prepared by:*
>
> *People Technology, Inc.*
> *2130 XXX Drive*
> *XXX, MI XXX*

PHASE I ENVIRONMENTAL SITE ASSESSMENT
OF SMITH INDUSTRIES
1111 ANYWHERE AVENUE
SOMEWHERE, MICHIGAN
August 1, 2000

Report Introduction

The introduction describes the scope and purpose, special term and conditions, limitations and exceptions of assessment, and limiting conditions and methodology used in the site assessment. Softshel, Inc. has an example of an introduction and a complete assessment template in their software package *Environmental Site Assessment v.4.0* (Refer to Example 3-2).

Example 3-2:

The purpose of this phase one environmental site assessment is to identify, within limits, the environmental liability, or potential environmental liability, of the property under assessment. Recognized environmental conditions will be assessed and reported herein.

Included in this report are aerial photography, an analysis of the land records of the site, interview and inspection results and findings and governmental regulatory records review as related to potential environmental liability.

It should be noted that when an assessment is completed without adequate subsurface exploration or chemical screening of soil and groundwater beneath the site, as in this study, no statement of scientific certainty could be made regarding latent subsurface conditions that may be the result of on-site or off-site sources. The findings and conclusions of this report are not scientific certainties, but rather, probabilities based on professional judgment concerning the significance of the data gathered during the course of the environmental assessment.

[Insert company or person] are not able to represent that the site or adjoining land contains no hazardous waste, oil or other latent condition beyond that detected or observed by this company during the assessment. The possibility always exists for contaminants to migrate through surface water, air or groundwater. The ability to accurately address the environmental risk associated with transport in these media is beyond the scope of this investigation.

EXECUTIVE SUMMARY

The executive summary is more detailed in describing the scope and purpose, special term and conditions, limitations and exceptions of assessment, and limiting conditions and methodology used in the site assessment than the introduction (Refer to Example 3-3).

The first paragraph of the executive summary shall contain the following:

- A sentence, which describes the name of the company who has completed the assessment and the location of the subject.
- A sentence, which describes the standards, used to conduct the assessment.
- A sentence, which describes who may rely on the assessment.

Example 3-3:

People Technology, Inc. (PT) has completed the Phase I Environmental Site Assessment (ESA) of an industrial building, located at 1111 Anywhere Avenue in Somewhere, XXX County, Michigan. The report was completed in accordance **with the minimum reporting requirements and procedures for environmental investigations for LENDING INSTITUTION-200x and LENDING INSTITUTION Phase I ESA Checklist dated January 1, 2000** and in general accordance with the American Society for Testing and Materials (ASTM) Standard Practice for Environmental Site Assessments: Phase I ESA Process (ASTM Designation: E-1527-00). It is a site-specific assessment that is related to the environmental conditions of the subject property only. Any additions or deletions from this practice are described in the Limitations Section. This study is a compilation of information obtained through visual inspection, inquiry into current and past ownership and uses of the subject property and review of standard environmental record sources. A site vicinity map is included as Section 11.2 and a site diagram is included as Section 11.3. **THIS REPORT WAS PREPARED FOR THE EXCLUSIVE USE OF MR. SMITH AND THE LENDING INSTITUTION, AND EACH MAY RELY ON THE REPORTS CONTENTS.**

OR,

People Technology, Inc., (PT) was retained by Mr. Joe Smith to conduct a Phase I Environmental Site Assessment (ESA) of located at 13/50 Somewhere Street, Somewhere, XXX County, Michigan. The report was completed in general accordance with the American Society for Testing and Materials (ASTM) Standard Practice for

Environmental Site Assessments: Phase I ESA Process (ASTM Designation: E-1527-00). Any additions or deletions from this practice are described in the Limitations Section. This study is a compilation of information obtained through visual inspection, inquiry into current and past ownership and uses of the subject property and review of standard environmental record sources.

Brief Description of the Subject property

This section describes the subject property (i.e. lot size, building, etc.) (Refer to Example 3-4).

Example 3-4:

The subject property is located at 1111 Anywhere Avenue and contains two one-story industrial buildings on approximately 1.18 acres of land. Throughout this Phase I ESA for historical reference the subject property's buildings will be referred to as 1111 Anywhere Avenue and 1112 Anywhere Avenue. According to the City of Somewhere Assessing records both buildings were built approximately in 1946. The building located at 1111 Anywhere Avenue is separated into the 1946 original building and has undergone building additions in 1950, 1952, 1957, and 1973. The current building is approximately 21,922 square feet in size, which is connected to municipal sewer and water, natural gas, and electricity.

The building located at 1112 Anywhere Avenue is separated into the 1946 original building and has undergone building additions in 1950, 1955, and 1956. The current building is approximately 10,314 square feet in size, which is connected to municipal sewer and water, natural gas, and electricity.

Historical Review Summary

This paragraph describes data failure and historical information, which were reasonably ascertainable during the record review of the subject property (Refer to Example 3-5).

Example 3-5:

Based on reasonably ascertainable records for the subject property, which extended back to approximately 1927, data failure occurred prior to 1927. The subject property, since building development approximately in 1946 the subject property has been occupied by two one-story buildings.

Site Reconnaissance

This paragraph explains the non-scope site reconnaissance findings that the client may want assessed in the assessment (Refer to Example 3-6).

Example 3-6:

During the July 9, 2000 site reconnaissance by a PT accredited asbestos building inspector of the subject property sampled suspect asbestos containing materials (ACM) building materials in the office area and in the 1946 original building of 1111 Anywhere Avenue. No suspect ACM was observed in the 1111 Anywhere Avenue building during the site reconnaissance. PT did not observe any sensitive ecological areas on the subject property, including potential wetlands.

Regulatory Review

This paragraph summarizes the regulatory database search performed on the subject property and its surrounding area (Refer to Example 3-7).

Example 3-7:

The regulatory records review identified one Large Quantity Generator, one Treatment, Storage, and Disposal Facility and one federal CERCLIS within state and federal regulatory records review as within the ASTM, approximate minimum search distance (AMSD) which represent a recognized environmental concern.

Data Failure

This section describes any data failure in connection with the subject property (Refer to Example 3-8).

ASTM E-1527 states a "standard historical source may be excluded (1) if the source is not reasonably ascertainable, or (2) if past experience indicates that the source is not likely to be sufficiently useful, accurate, or complete in terms of satisfying the uses of the property."

Example 3-8:

Based on reasonably ascertainable records for the subject property which extended back to approximately 1927, data failure occurred prior to that date and the following dates 1928 to 1934, 1936 to 1940, 1942 to 1949, 1951 to 1953, 1955 to 1962, 1964, 1965, 1967, 1969 to 1972, 1974, 1976, 1978, 1979, 1981, 1983, 1985, 1986, 1988, 1989, 1991, 1993, 1994, and 1996. The subject property, since building development in approximately in 1946 has been occupied by two one-story buildings. City directories of 1927, 1935, and 1941 had no listings for the subject property.

Findings and Conclusions

This paragraph contains the professional opinion of the custodian's findings and conclusions. If recognized environmental conditions were revealed during the site reconnaissance this section would describe the location and migration pathways (i.e. cracks in concrete or asphalt, drains, exposed soil, streams, rivers, pits, sumps, ponds, etc.) of them (Refer to Example 3–9).

Example 3-9:

It is the professional opinion of PT, an appropriate level of inquiry has been made into the previous ownership and uses of the property consistent with good commercial and customary practice in an effort to minimize liability, and no evidence or indication of Recognized Environmental Conditions has been revealed, except for the following:

- The original building section (1946 construction) located at 1111 Anywhere Avenue currently contains approximately two 1,000 gallons emergency open aboveground steel holding tanks which contain sludge and waste water which is generated at the power wash station. The sludge and wastewater contents have the potential to contain (lead, chromium, and other metals). The sludge and wastewater inside these tanks are held on-site for a period of up to one year and the contents are pumped into a cargo tank truck for off-site transport. Due to the long-term operations of the subject property as a paint stripping facility the potential exists for sludge contaminates to spill, leak, and overflow into cracks in the concrete floor and migrate into the soil and groundwater. Thus, these tanks represent a recognized environmental concern to the subject property

- The north wall of the original building section (1946 construction) located at 1111 Anywhere Avenue once contained a former paint booth. Historic records review indicated that the paint booth was operational from at least 1981 to 1995. Due to past long term operations of this paint booth there is the potential for releases of solvents and other contaminants. These solvents and other contaminants are contained in the paints such as chromium, lead, and other metals, which can migrate into the soil, and groundwater through cracks in the concrete floor and represents a recognized environmental concern.

- The south adjoining property historically has been a lithographic and printing facility from at least 1949 to 1968. Printing and lithographic facilities utilize printing fluids, inks, and solvents in there printing process. Past hazardous waste management practices for these wastes might have involved disposing of the waste on-site (i.e. dumping the waste out a side door onto the dirt ground) and this waste has the potential to migrate into the soil and groundwater of the subject property. Thus, this represents a recognized environmental concern.

- The east adjoining property is listed on the U.S. EPA CERCLIS list according to the VISTA database search. There is the potential for off-site migration of contaminates to the subject property. PT has FOIA the US EPA Region 5 District Office regarding the file on the site until PT receives further information regarding this site it represents a recognized environmental concern to the subject property.

- Two chemical stripping tanks are located against the northern wall in the 1957 building addition of 1111 Anywhere Avenue. Each tank contains approximately 240 gallons of HCL and NaOH solution. It appears that residue from HCL and NaOH covers the concrete floor and there is also evidence that both tanks are stained and rusting. The HCL and NaOH are both corrosive chemicals, which have the potential to erode the concrete floor thus, allowing HCL and NaOH to migrate into the soil and groundwater. Therefore, this area represents a recognized environmental concern.

- The storage area behind the 1112 Anywhere Avenue historically and currently has been utilized as a storage area for metal parts, metal shavings, and metal 55-gallon drums. Metal shavings are a by-product of metal fabricators facilities, which usually contain cutting oils and other solvents. The metal 55-gallon drums may have contained petroleum products which had the potential to leak, and spill unto the dirt ground and the severely damage asphalt pavement. Thus, these contaminates have the potential to migrate into the soil and groundwater which represents a recognized environmental concern.

- The 1111 Anywhere Avenue building according to historical records once contained a metal cutting and forming operation, and a glass working plant. Metal fabricators and glass works operations usually use cutting oils, solvents, cleaners, resins, other petroleum products, and metals in their processes. These contaminates have the potential to migrate into the soil and groundwater through cracks in the concrete floor and represents a recognized environmental concern.

- There appears to be a lack of proper spill prevention practices located in the 1950 building addition of the 1111 Anywhere Avenue building. This area contains one 55-gallon drum of HCL and one 55-gallon drum of NaOH. There is evidence of recent spillage from both drums onto the concrete floor. The potential exists for HCL and NaOH to migrate into the soil and groundwater through cracks in the concrete floor and represents a recognized environmental concern.

- The 1112 Anywhere Avenue building has a stained floor, which appears to be from a spill of an unknown substance. The floor slopes toward the sewer drain located in the southwestern bay. Contaminates have the potential to migrate into the soils and groundwater from cracks in the concrete which represents a recognized environmental concern.

- The City of Somewhere Assessor's records indicated that the 1111 Anywhere Avenue building once had a boiler there is no records to indicate if a tank exists or have been removed from the subject property. There also appears to be an indication that the 1952 building addition was heated by forced air-oil. Boilers usually utilize oil, as a fuel source and therefore, an UST could still be present on the subject property. The UST has the potential to leak and their contents would have the potential to migrate into the soil and groundwater. Thus, this represents a recognized environmental concern.

- The 1111 Anywhere Avenue building has been associated with the Smith Hardware Manufacturing Company in 1954 and as a vehicle maintenance garage according to the City of Somewhere Assessing records. The hardware manufacturers utilize metals, cutting oils and other solvents in the producing their products. Vehicle maintenance garages have been associated with petroleum products and solvents. The potential exists for these contaminates to migrate into soil and groundwater from cracks in the concrete floor caused by the lack of proper spill control and management of these materials and thus, represents a recognized environmental concern.

Table Of Contents

This section lists the structure and contents of the assessment report (Refer to Example 3-10).

Example 3-10:

INTRODUCTION
EXECUTIVE SUMMARY
LIMITATIONS

SECTION 1.0—HISTORICAL REVIEW:
Section 1.1: Chain of Title
Section 1.1a: Environmental Liens
Section 1.2: Aerial Photographs
Section 1.3: Fire Insurance Maps
Section 1.4: City Directories
Section 1.5: Interviews
Section 1.6: Property Tax Records
Section 1.7: Previous Site Investigations
Section 1.8: Other Sources/Historical Plat Maps

SECTION 2.0—REGULATORY FILE REVIEW:
Section 2.1 NPL
Section 2.2 CERCLIS
Section 2.3 State Master Lists
Section 2.4 TSDFs
Section 2.5 RCRA Generators
Section 2.6 ERNS
Section 2.7 LUSTs
Section 2.8 Solid Waste Landfills
Section 2.9 USTs
Section 2.9.1 ODNR
Section 2.10 National Heritage Program
Section 2.11 State Historical Society

Section 2.12 Local Fire Department
Section 2.13 Local Health Department
Section 2.14 Local Building Department
Section 2.15 Local Zoning Department
Section 2.16 Map

SECTION 3.0—SITE INSPECTION:
Section 3.1.0: Name, Date, and Time of Inspection
Section 3.1.1: Weather
Section 3.1.2: Physical and Visual Observations
Section 3.1.3: Current Property Use
Section 3.1.4: Intended Property Use
Section 3.1.5: Utilities
Section 3.2: Water Wells
Section 3.2.1: Sanitary Sewer/Septic System
Section 3.2.2: Interviews with Knowledgeable Site Contacts
Section 3.3: Buildings. Interior
Section 3.3.1: Interior Waste Streams
Section 3.4: Building Exterior
Section 3.4.1: Exterior Site Operations
Section 3.4.2: Hazardous Waste Disposal
Section 3.4.2.a: Agricultural Waste
Section 3.4.3: Solid Waste Stream and Disposal
Section 3.5: Underground and Above Ground Storage Tanks
Section 3.5.1: Pits, Ponds, Lagoons and Waste Disposal Area
Section 3.6: Polychlorinated Biphenyls
Section 3.6.1: Electromagnetic Fields
Section 3.7: Vegetation
Section 3.8: Oil and Gas Wells
Section 3.9: Topography
Section 3.9.1: Drainage Patterns
Section 3.9.2: General Soil Profile
Section 3.9.3: Geology
Section 3.9.4: Local Groundwater Flow
Section 3.9.5: Legal Description of Property
Section 3.10.0: Adjoining Properties
Section 3.10.1: North Adjoining Property
Section 3.10.2: South Adjoining Property
Section 3.10.3: East Adjoining Property
Section 3.10.4: West Adjoining Property

NON-SCOPE AND ADDITIONAL SOURCES
SECTION 4.0—ASBESTOS CONTAINING MATERIALS (ACM) INSPECTION
Section 4.1: Identification of Certified Asbestos Inspector
Section 4.2: Suspect Non-Friable ACM in Good Condition
Section 4.3: Suspect Non-Friable ACM in Damaged Condition
Section 4.4: Analytical Results of Suspect Non-Friable ACM
Section 4.5: Suspect Friable ACM in Good Condition
Section 4.6: Analytical Results of Suspect Friable ACM
Section 4.7: Suspect Friable ACM in Damaged Condition
Section 4.8: Analytical Results of Suspect Friable ACM (Damaged)

SECTION 5.0—LEAD BASE PAINT SCREEN
SECTION 6.0—PRELIMINARY RADON INSPECTION
SECTION 7.0—LEAD IN WATER TESTING
SECTION 8.0—WETLANDS INVESTIGATION
SECTION 9.0—MULTI-MEDIA COMPLIANCE INSPECTION
Section 9.1.1 Air Emissions
Section 9.1.2 Waste Water Discharge
Section 9.1.3 RCRA Waste Generate and Disposal

SECTION 10.0—CONCLUSIONS AND RECOMMENDATIONS

Section 10.1 Conclusions for the Phase I ESA
Section 10.2 Recommendation for the Phase I ESA

SECTION 11.0—APPENDICES
Section 11.1: Documentation of Sources
Section 11.2: Map of Subject Property Location
Section 11.3: Drawing of Subject Property
Section 11.4: Color Photos of the Subject Property
Section 11.5: USGS 7.5 Quad Map
Section 11.6: Aerial Photographs
Section 11.7: Fire Insurance Maps
Section 11.8: Regulatory File Review Documentation
Section 11.9: County Soil Survey Soil Map
Section 11.10 Pertinent Well Log Information
Section 11.11: Legal Description of Subject Property
Section 11.12: Property Tax File Documentation
Section 11.13 City Directory Documentation

Section 11.14: FOIA Request/Responses
Section 11.15: Any Previous Reports
Section 11.16: ACM inspector certificate/chain of custody/analytical results
Section 11.17: LBP inspector certificate/chain of custody/analytical results
Section 11.18: Radon inspector certificate/chain of custody/analytical results
Section 11.19: Lead in Water Sample Chain of Custody/analytical results
Section 11.20: Supplemental Information
Section 11.21: Site Custodian's Resume

Limitations

The environmental professional shall document, in the report, general limitations and bases of review, including limitations imposed by physical obstructions such as adjacent buildings, bodies of water, asphalt, or other paved areas, and limiting conditions (for example, snow, rain).

Preparation

This paragraph describes what standards and guidelines were used in the assessment. They consist of the American Society for Testing and Materials (ASTM) Standard Practice for Environmental Site Assessments: Phase I ESA Process (ASTM Designation: E-1527-00) and/or any additional lending institution guidelines.

Example 3-11:

This Phase I Environmental Site Assessment (ESA) was prepared in accordance with the minimum reporting requirements and procedures for environmental investigations for LENDING INSTITUTION-200x and LENDING INSTITUTION Phase I ESA Checklist dated January 1, 20xx and in general accordance with the American Society for Testing and Materials (ASTM) Standard Practice for Environmental Site Assessments: Phase I ESA Process (ASTM Designation: E-1527-00). It is a site-specific assessment that is related to the environmental conditions of the subject property only.

Additions or Deletions

This section describes any deletions and deviations from the ASTM E-1527 practice.

Example 3-12:

There are no deletions to the ASTM Standard. Additions to the ASTM standard include LENDING INSTITUTION's requirements for: visual inspection and limited bulk asbestos containing materials (ACMs) (i.e. friable and non-friable); requirement for the environmental professional to check federal, state, and local cleanup liens against the subject property.

OR,

Deletions to the ASTM Standard include a formal site reconnaissance with access to all portions of the interior was granted, as well as, the current subject property owner was reluctant to provide any information related to the subject property. Additions to the ASTM standard include LENDING INSTITUTION's requirements for: requirement for the environmental professional to check federal, state, and local cleanup liens against the subject property; expansion of the ASTM Approximate Minimum Search Distance (AMSD) for CERCLIS sites to 1-mile, ERNS sites to 1/4 mile, Solid Waste Landfills to 1-mile.

Warranties, Guarantees, and Certification

The environmental professional will probably not certify (i.e., warrant) that a site is free of recognized environmental conditions, because it is impossible to know if such a condition exists or does not exists.

Areas that were not surveyed or sampled may contain recognized environmental conditions. These recognized environmental conditions may have migrated to areas that showed no signs of contamination when sampled before. By giving the subject property a certification that certain conditions exist opens the door for a negligent ruling in the courts. Thus, the environmental professional can only provide their professional opinion.

The contract and the report should include wording which makes the client responsible for project related liabilities that consultants have no control (Refer to Example 3-13).

Example 3-13:

In preparing the assessment report, People Technology, Inc., may have relied on information obtained from or provided by others. People Technology, Inc., makes no representation or warranty regarding the accuracy or completeness of this information gathered through outside sources or subcontracted services. No warranty, guarantee, or certification of any kind, expressed or implied, at common law or created by statute, is extended, made, or intended by rendering these environmental consulting services or by furnishing this written report. Environmental conditions and regulations are subject to constant change and reinterpretation. One should not assume that any on-site conditions and/or regulatory statutes or rules would remain constant in the future, after People Technology, Inc., has completed the scope of work for this project. Furthermore, because of the facts stated in this report are subject to professional interpretation, other professionals could reach differing conclusions.

The Phase I ESA was performed utilizing American Society for Testing and Materials (ASTM) Designation E-1527-00. The Phase I ESA did not include any additions or deletions from the ASTM standard. Appendix A includes documentation of sources checked as part of the Phase I ESA scope of work.

Contaminants may be hidden in subsurface material, covered by pavement, vegetation, or other substances. Additionally, contamination may not be present in predictable locations. The most that People Technology, Inc., can do is prepare a logical assessment program to reduce the client's risk of discovering unknown contamination. This risk may be reduced by more extensive exploration on the site. Even with additional exploration, it is not possible to completely eliminate the risk of discovering contamination on-site. It cannot be assumed that samples collected and conditions observed are representative of an area that has not been sampled and/or tested. Tests and other data collected for the report were obtained only for the sole purposes stated in this report, and they should not be used for purposes or reasons other than those intended.

Present on the subject property were areas of concrete and asphalt covered surfaces, building structures, and interior storage conditions which limited PT's observation of the ground surface. People Technology, Inc., cannot make any conclusions regarding the potential for visible signs of contamination at areas covered by the aforementioned items. People Technology, Inc., can further evaluate these areas at a later date when the land/surface cover is not present.

ASTM Standard Practice E-1527-00 defines "all appropriate inquiry." ASTM's definition is aimed at providing an industry standard in an effort to guide legal interpretation of the Comprehensive Environmental Response, Compensation and Liability Act's (CERCLA's) "innocent land owner defense." Even with ASTM's definition, the level of investigation necessary to demonstrate "due diligence" or "all appropriate inquiry" has not been legislatively defined by federal, state, or local law; therefore, People Technology, Inc., cannot warrant that the work undertaken for this report will satisfy "due diligence," "all appropriate inquiry," or any other similar standard under any federal, state, or local law.

Due to changing environmental regulatory conditions and potential on-site activities after the assessment, the client may rely upon the conclusions within this assessment for a period of 180-days from the report's issuance date.

Any reports, field data, field notes, laboratory testing, calculations, estimates or other documents prepared by or relied upon by People Technology, Inc., are the property of People Technology, Inc. If any of these documents are released or obtained by a party other than the client, People Technology, Inc., may not discuss the project with that party unless the original contracted client notifies People Technology, Inc., of the same and People Technology, Inc. is authorized to disclose the information and to discuss the project with others. Except as otherwise agreed with the client, People Technology, Inc. further states that it disclaims any duty of any kind or nature to any person or entity other than the client in preparing this report.

People Technology, Inc., does not assume liability for any losses or damages that the client or third party incur due to the results or conclusions provided in this assessment.

Another resource pertaining to the certification and statement of limiting conditions can be obtained from the Environmental Assessment Association (Refer to Example 3-14).

Example 3-14:

CERTIFICATION AND STATEMENT OF LIMITING CONDITIONS

Certification: The Environmental Inspector (Inspector) certifies to the Buyer, Seller and/or lender in a transaction as named in the inspection report "Principal Parties"; and the Inspector and Principal Parties agree that:

1. The Environmental Inspector has no present or contemplated future (a) partnership with Principal Parties nor (b) an interest in the property inspected which could adversely affect the inspector's ability to perform an objective inspection; and

neither the employment of the inspector to conduct the inspection, nor the compensation for it, is contingent on the results of the inspection.

2. The Environmental Inspector has no personal interest in or bias with respect to the subject matter of the inspection report or any parties who may be part of a financial transaction involving the property. The conclusions and recommendations of the report are not based in whole or in part upon the race, color, creed, sex or national origin of any of the Principal Parties.

3. The Environmental Inspector has personally inspected the property, both inside and out and has made visual inspection of adjacent properties, to the extent possible by readily available access. The inspection does not include the removal of any soil, water or air samples, the moving of furniture or fixtures, or any type of inspection that would require extraordinary effort to access.

4. All contingent and limiting conditions are contained herein (imposed by the terms of the inspection assignment or by the undersigned affecting the conclusions and recommendations contained in the report).

5. This Environmental Inspection report has been made in conformity with and is subject to the requirements of the Code of Professional Ethics of the Environmental Assessment Association.

6. All opinions, conclusions and recommendations concerning the inspected property that are set forth in the inspection report were prepared by the Environmental Inspector whose signature appears on the report. No change of any item in the report shall be made by anyone other than the Inspector, and the Inspector shall have no responsibility for any such unauthorized change.

Contingent and Limiting Conditions: The certification of the Environmental Inspector appearing in the environmental inspection report is subject to the following conditions and to such other specific and limiting conditions as are set forth by the Inspector in the report.

1. The Inspector assumes no responsibility for matters of a legal nature affecting the property inspected or the title thereto. The property is inspected assuming responsible ownership.

2. Any sketch appearing in or attached to the inspection report, or any statement of dimensions, capacities, quantities or distances, are approximate and are included to assist the reader in visualizing the property. The Inspector has made no survey of the property.

3. The Inspector is required to give testimony or appear in court because of having made the inspection with reference to the property in question, unless arrangements have been previously made therefore.

4. This report is not intended to have any direct effect on the value of the property inspected but simply to provide a visual Environmental Assessment solely for the benefit of the Principal Parties.

5. The Inspector assumes that there are no hidden, unapparent, or latent conditions or defects in or of the property, subsoil, or structures, other than those noted on the inspection report or any addendum to the report which the Inspector assumes nor responsibility for such conditions, or for the inspection, engineering or repair which might be required to discover or correct such factors.

6. Information, estimates and opinions furnished to the Inspector, and contained in the report, were obtained from sources considered reliable and believed to be true and correct. However, the Inspector has made no independent investigation as to such matters and undertakes no responsibility for the accuracy of such items.

7. The Inspection and Inspection Report are made by the Inspector solely for the benefit and personal use of the Principal Parties. Disclosure of the contents of the Inspection Report is governed by the Bylaws and Regulations of the Environmental Assessment Association. No disclosure may be made of the Inspection Report without prior written consent of the Inspector and the Inspector undertakes no responsibility for harm or damages to any party other than the Principal Parties.

8. Neither the Inspection report, any part thereof, nor any copy of the same (including conclusions or recommendations, the identity of the Inspector, professional designation, reference to any professional organization, or the firm with which the Inspector is connected), shall be used for any purposes by anyone but the Principal Parties. The report shall not be conveyed by anyone to the public through advertising, public relations, news, sales, or other media, without the prior written consent and approval of the Inspector.

Chapter 4: From Historical Record Reviews to Map of the Subject Property

The topics discussed in this chapter include:
- Historical Records Review
- Site Reconnaissance
- Building Interior
- Building Exterior
- Legal Description of the Property
- Adjoining Properties
- Regulatory File Review

HISTORICAL RECORDS REVIEW

A historical records review must be conducted for sites developed from 1940 to the present, standard historical sources must be reviewed. For sites with development prior to 1940, at least one standard source must be reviewed to the time when no development was present.

To determine if the subject property has the potential to have hazardous material and petroleum products contamination, a historical records review for the subject property must be completed. This is accomplished by evaluating the current and past uses of the subject property. The research must be completed to determine whether operations were conducted that involved the use, storage and/or disposal of hazardous waste, hazardous substances, and/or petroleum products on the subject property (Refer to Example 4-1).

Example 4-1:

An understanding of the subject property area was obtained from reasonably ascertainable standard and other historical sources extending back to 1927. Standard and other historical sources were able to document the first industrial developed use of the subject property as an industrial building. The subject property buildings were constructed in approximately in 1946. Data failure occurred prior to 1927 the following dates 1928 to 1934, 1936 to 1940, 1942 to 1949, 1951 to 1953, 1955 to 1962, 1964, 1965, 1967, 1969 to 1972, 1974, 1976, 1978, 1979, 1981, 1983, 1985, 1986, 1988, 1989, 1991, 1993, 1994, and 1996 for which there were no reasonably ascertainable standard historical sources or interviewees with knowledge during these time periods. Interviewees provided independent knowledge of subject property and surrounding area usage which in turn provided information confirming historical subject property and general adjoining and surrounding land usage. See Section 11.1 for a Bibliography for specific documentation of standard and other historical sources consulted and availability of these sources. The history of the subject property and adjoining and surrounding areas, which was able to be derived from standard historical sources and other sources to satisfy the ASTM standard requirements for uses of a property (except those excluded by data failure), have been described within the text of this report.

The subject property located at 1111 Anywhere Avenue contains two one-story industrial buildings that are approximately 1.18 acres in size. The frontage of the subject property is approximately 187 feet along Anywhere Avenue Street and extends north approximately 278 feet to the fence line.

For historical reference these buildings will be referred to as 1111 Anywhere Avenue and 1112 Anywhere Avenue throughout the Phase I ESA. Both buildings were built in approximately in 1946 according to the City of Somewhere Assessing records. The building located at 1111 Anywhere Avenue is separated into the 1946 original building and has undergone building additions in 1950, 1952, 1957, and 1973. The current building is approximately 21,922 square feet in size that is connected to municipal sewer and water, natural gas, and electricity.

The building located at 1112 Anywhere Avenue is separated into the 1946 original building and has undergone building additions in 1950, 1955, and 1956. The current building is approximately 10,314 square feet in size that is connected to municipal sewer and water, natural gas, and electricity.

Since 1946 two industrial buildings that were utilized by the Smith Glass Corporation, a glass washing facility, and Smith Associates, a metal fabricator, have occupied the subject property. The current usage of the subject property consists of Smith Industrial Services a painted parts stripping, sandblasting, and painting industrial service company.

Chain of Title Search

A chain of title is also referred to, as a title record is one standard source of identifying previous ownership for subject property especially if development occurred prior 1940.

Depending on time, resources, and your client a title search may be conducted for the subject property. However, sufficient previous ownership, users, and occupants can be identified through other review of reasonably ascertainable sources (Refer to Example 4-2).

Example 4-2:

Based upon PT's professional opinion, the documented history of the subject property indicates it was originally developed for industrial usage since 1946 and currently is occupied by two one-story industrial buildings. Sufficient previous ownership, users, and occupants have been identified through review of reasonably ascertainable sources.

Information Regarding Environmental Liens

The user must disclose any information pertaining to existing and/or past environmental liens in connection with the subject property.

Aerial Photographs

Aerial photographs are preformed by various United State agencies and private companies the oldest photographs date back to the middle 1930s.

Historical aerial photography is often useful in identifying past usages of a property or surrounding area, building locations, and discernible notable features, which may help identify past and current recognized environmental concerns in connection with the subject property and surrounding area (Refer to Example 4-3).

Example 4-3:

Reasonably ascertainable aerial photography was reviewed for 1966 (scale: 1":1,000'), 1975 (scale: 1":1,000'), 1980 (scale: 1":1,000') and 1995 (scale: 1":1,000'). Aerial photographs from 1966, 1975, 1980, and 1995 were obtained from the XXX (XXX), Somewhere, Michigan. Photocopied reproductions of these aerials have been included in Section 11.6. Reference to the subject property or adjoining site usages in a particular aerial year is based on information obtained through site observations and standard and other historical sources. It should be noted that the scale and resolution of aerial photographs provided for only general descriptions of subject property and adjoining properties and limited description and discernment of site-specific features. Refer to Section 11.6 for photocopies of aerial photographs.

Data failure occurred for aerial photographs during the years 1940's and 1950's.

1966

The subject property appears to be utilized as an industrial building. The north property contains railroad tracks and beyond is the north adjoining property appears to be utilized as an industrial building. The adjoining west property appears to be utilized as a truck yard. The south adjoining property appears to be utilized as an industrial building. The east adjoining property appears to be utilized as an industrial building.

1975

The subject property appears to be utilized as an industrial building and appears to have added a truck well and a building addition. The subject property buildings appear to be similar in layout and location as observed during the July 9, 2000 site reconnaissance. The north property contains railroad tracks and beyond is the north adjoining property appears to be utilized as an industrial building. The adjoining west property appears to be utilized as a truck yard. The south adjoining property appears to be utilized as an industrial building. The east adjoining property appears to be utilized as an industrial building.

1980

The subject property appears to be utilized as an industrial building. The subject property building appears to be similar in layout and location as observed during the July 9, 2000 site reconnaissance. The north property contains railroad tracks and beyond is the north adjoining property appears to be utilized as an industrial building. The adjoining west property appears to be a vacant lot. The south adjoining property appears to be utilized as an industrial building. The east adjoining property appears to be utilized as an industrial building.

1995

The aerial photograph appearance is the same as 1980.

Fire Insurance Maps

Fire insurance maps were initially produced for the insurance industry to provide information on the fire risks of buildings and other structures (Refer to Example 4-4). Some of the publishers include:

- Sanborn
- Nathan Nirenstein
- Perris
- Spielman and Brush
- Hexamer
- Scarlett
- The Fire Underwriters Inspection Bureau.

The most well known publisher from the list above is the Sanborn Fire Insurance Maps. Sanborn Fire Insurance Maps typically are dated from the late 1800's to the 1950's, and include updates for selected areas as recently as 1990.

There are approximately 12,000 United State cities and towns with fire insurance maps from the period 1852 to the present.

These maps often provide data that contain information used to determine the presence of underground and aboveground storage tanks, type of building materials, location of flammable material storage, and types of businesses that occupied a particular site.

Example 4-4:

Based on review with the Library, there were Sanborn Fire Insurance Maps available for the subject property. Refer to Section 11.7 for photocopies of the following years:

1950

The subject property appears to be utilized as two separate one-story industrial buildings. One building is located at 1111 Anywhere Avenue that appears to be utilized by the Smith Glass Company. The other building located 1112 Anywhere Avenue appears to be utilized as a factory. The east adjoining property appears to be a vacant lot. The south adjoining property at 1114 Anywhere Avenue appears to be utilized by the XXX Arts Process Company, a lithographing shop. The west adjoining property appears to be utilized by the XXX Chrome and Chemical Company, a chemical plant. There are no indication on the Sanborn Map of any above ground or underground storage tanks on the subject property or adjoining properties.

1984

The subject property appears to be utilized as two separate one-story industrial buildings. The subject property appears to be utilized as two separate one-story buildings. One building is located at 1111 Anywhere Avenue appears to be utilized by the Smith Glass Company. The building is the same as the 1950 Sanborn Map with exception of the 1955 and 1957 building additions added to 1111 Anywhere Avenue. The east adjoining property appears to be vacant land. The south adjoining property at 9751 Anywhere Avenue appears to be utilized by the XXX Arts Process Company, a lithographing shop. The west adjoining property appears to be utilized by the XXX Chrome and Chemical Company, a chemical plant. There are no indication on the Sanborn Map of any above ground or underground storage tanks on the subject property or adjoining properties.

1987

The Sanborn Map has not changed since 1984.

1990

The Sanborn Map has not changed since 1987.

City Directories

City directories have been published for cities and towns across the United States since the 1700's. Originally they were a list of town residents then the city directories developed into a reference tool for locating individuals and businesses in a particular urban or suburban area.

City directories are published annually by companies such as Bresser's Cross Index City Directories, and R.L. Polk's Directories, and may be located in city and university libraries (Refer to Example 4-5).

Example 4-5:

A review of reasonably ascertainable Bresser's Cross-Index City Directories, and any other reasonably ascertainable city directory coverage was completed by PT. Bresser's Cross-Index City Directories were available for the subject property area between 1973 and 1997. R. L. Polk's Directories were available for the subject property area between 1927 and 1968. It should not be construed that the earliest date represented is the initial date of occupancy. Other descriptions regarding adjoining property, historical usage, and development should also be referenced as provided in the text of this report. These directories provided the following limited listings of the subject property and sites adjoining and surrounding the subject property. Refer to Section 11.13 for photocopies of city directories.

1111 Anywhere Avenue Street (Subject property)

1997:	Smith Industrial Services
1992:	Smith Associates/Smith Engineers
1987:	SmithAssociates
1982:	Smith Glass Corp.
1977:	Smith Glass Corp.
1973:	Smith Glass Corp.
1968:	Smith Glass Corp.
1963:	Smith Glass Corp.
1954:	Smith Glass Corp.
1941:	Not Listed
1935:	Not Listed
1927:	Not Listed

1112 Anywhere Avenue (Subject property)

1997:	Not Listed
1992:	Not Listed
1987:	Not Listed
1982:	Not Listed
1977:	Not Listed
1973:	Not Listed
1968:	Not Listed
1963:	Not Listed
1954:	Smith Hardware Manufacturing Company
1941:	Not Listed
1935:	Not Listed
1927:	Not Listed

(West Adjoining Property)

The west adjoining property is a vacant lot. However, past aerial photographs indicate that it might have been utilized as a truck or automobile lot. Approximately 0.18 mile west of the property is the closest west adjoining property which is located at the corner of ABC and XYZ.

9701 XYZ Avenue (West Adjoining Property)

1997:	XXX Express Ltd
1992:	XXX American Gas Service
1987:	XXX American Gas Service
1982:	XYZ ABC Gulf
1977:	Not Listed
1973:	XYZ and ABC Gas Service
1968:	XYZ and ABC Gas Station
1963:	XXX Auto Sales
1954:	XXX Motor Sales
1941:	Not Listed
1935:	Not Listed
1927:	Not Listed

8611 ABC (East Adjoining Property)

1997:	Not Listed
1992:	Not Listed
1987:	Not Listed
1982:	Not Listed
1977:	Not Listed
1973:	Not Listed
1968:	Not Listed
1963:	Not Listed
1954:	Not Listed
1941:	Not Listed
1935:	Not Listed
1927:	Not Listed

1114 Anywhere Avenue (South Adjoining Property)

1997:	XXX Architecture Materials
1992:	XXX Architecture Materials
1987:	XXX Company/XXX Housekeeper
1982:	XXX/XXX
1977:	XXX/XXX
1973:	XXX/XXX
1968:	XXX Printing and Lithographing Company
1963:	XXX Arts Process Company
1954:	XXX Arts Process Company
1941:	Not Listed
1935:	Not Listed
1927:	Not Listed

Interviews

The interview is one key component of a site assessment and is discussed in detailed in Chapter 2 (Refer to Example 4-6).

Example 4-6:

On July 9, 2000 an interview was conducted by PT with Ms. Mary Smith of Smith Industrial Services, Somewhere, Michigan a secretary representing the subject property owner. Ms. Smith did not indicate to PT any knowledge of environmental liens encumbering the subject property, or any pending, threatened, or past environmental litigation, environmental administrative procedures, or notices from government entities regarding possible violations of environmental law or possible environmental liability.

Documentation of individuals interviewed or attempts to interview individuals is included in Section 11.1. In general the interviewees provided information consistent with that identified by other sources consulted and referenced in Section 11.1.

Interviews with Knowledgeable Site Contacts

As describe in Chapter 2, the owner, user, the key site manager, and one staff official from one local or state agency must be interviewed regarding the subject property (Refer to Example 4-7).

Example 4-7:

PT contacted the City of Somewhere Offices, City of Somewhere Health Division, which is identified in Section 11.1. In general, the interviewee provided information consistent with other sources consulted in Section 11.1 and observations made during the July 9, 2000 site reconnaissance.

Property Tax Records

The building and assessor records provide information about how the subject property is zoned and how the property evolved from undeveloped land to its current use.

The assessor's records usually contain the property tax file that includes records of past ownership, appraisals, maps, sketches, photos, or other information of the subject property.

These records might indicate any installation or removal of aboveground and underground storage tanks on the subject property (Refer to Example 4-8).

Example 4-8:

The City of Somewhere Assessor Records, Somewhere, Michigan, has the subject property currently zoned as M-2, Restricted Light Industrial and since 1946 two industrial buildings were located on the subject property.

Reasonably ascertainable assessment information provided through the City of Somewhere Assessor's Office, Somewhere, Michigan, was obtained and reviewed. Copies of the current and oldest assessment records for the subject property, which were reasonably ascertainable, are included in Section 11.12.

Previous Site Investigations

This section describes a summary of earlier reports in connection with the subject property. A report might conclude, that no further action to be taken on the subject property at this time or list recognized environmental concerns in connection with the subject property (Refer to Example 4-9). The ASTM E-1527 practice states "standard historical sources reviewed as part of a prior environmental site assessment do not need to be searched for or reviewed again, but uses of the property since the prior environmental site assessment should be identified either through standard historical sources or by alternatives to standard historical sources, to the extent such information is reasonably ascertainable."

Example 4-9:

PT obtained a previous Phase I ESA preformed by XXX Environmental Inc., Birmingham, Michigan, on April 5, 1995 refer to Section 11.15.

The Phase I ESA concluded that 1111 Anywhere Avenue had been utilized for various metal cutting and forming operations. The Phase I ESA indicated the areas inside the building included stained concrete flooring within the active process area. This staining was noted to be the result of incidental spills of cutting fluids and other liquids associated with metal cutting operations. The Phase I ESA also indicated extensive paint dust and residue were observed within the spray paint booth area.

The 1112 Anywhere Avenue building has a history of manufacturing and of a vehicle maintenance garage operation. The Phase I ESA There was no evidence of petroleum storage currently at this location. A hydraulic hoist was identified and was addressed in the September 1996 Phase II ESA.

The surrounding properties of the subject property discovered the following regulatory file review identified the east adjoining property as being a CERCLIS site of documented or suspected contamination has the potential to migrate to the subject property and is a recognized environmental concern. The south adjoining property has been identified as being a printing facility and a LUST site. Printing facilities have been associated with solvents and inks thus; this site is a recognized environmental concern. The LUST for this site has been addressed in Section 4.7 of this report and is not a recognized environmental concern.

The previous owner and key-site manager, Mr. John Brown of Smith Associates, a previous occupant and Mr. Jon Jones of Smith Glass Corporation, previous owner of the property were interviewed in the 1995 Phase I ESA.

Mr. Brown stated that Smith Associates had operated as a metal forming and welding operation in 1111 Anywhere Avenue building. The facility during their process utilized various cutting fluids and operated under a Wayne County Department of Environment air permit for the spray paint booth. Mr. Brown indicated that he was not aware of the site having underground storage tanks for chemical or petroleum storage.

The interview with Mr. Jones indicated that the 1112 Anywhere Avenue remained a warehouse and/or vacant from the time he purchased it in 1981 to 1995. The 1111 Anywhere Avenue building has operated as the Smith Glass Corporation from 1947 to 1981. The Smith Glass Corporation process included an industrial-sized washing machine that was utilized to clean glass products. Wastewater from this process was reportedly discharged into the sanitary sewer servicing the site. Mr. Brown indicated that he was not aware of the site having USTs for chemical or petroleum storage.

Smith Consulting and Technology, Inc. (SCT), Somewhere, Michigan in September of 1996, conducted a previous Phase II ESA on the subject property which was performed. Five hand auger borings (HA-1 through HA-5) were performed at the subject property location to address recognized environmental concerns identified in the April 5, 1995 Phase I ESA performed by Smith Environmental, Inc., and also recognized environmental concerns observed by SCT during an initial site-visit. Three borings (HA-2 through HA-4) were performed inside the 1112 building, one boring (HA-1) was performed outside the west end of the building, and one boring (HA-5) was performed inside the 9797 building. Soil samples were collected for laboratory analysis at depths of two to three feet below ground surface (bgs), two and one-half to three feet bgs, and three to four feet bgs. These locations were selected as the mostly likely zones of contamination based upon visual observations. Samples HA-2 (2-3') and HA-3 (2.5-3') were composited due to the typical chemical composition of hydraulic oil and submitted for laboratory analysis for polychlorinated biphenyls (PCBs). Individual soil samples were also collected from HA-2 and HA-3 and submitted for benzene, toluene, ethylbenzene, total xylenes (BTEX) and polynuclear aromatics (PNA) analysis. HA-1 (2-3') and HA-4 (2-3') were submitted for BTEX and PNAs, while HA-5 (3-4') was submitted for a volatile organic compound scan and PNAs.

Analytical results that contained concentrations above the method detection limits are summarized in Table 1, Soil Analytical Results. Analytical results demonstrate detectable concentrations of concern within HA-1 through HA-4. Concentrations within soil sample HA-5 were all below method detection limits. Analytical results for all soil samples collected are below the residential, soil leaching to ground water contaminant criteria for soil outlined in the Michigan Department of Environmental Quality—Environmental Response Division (ERD) Operational Memorandum Number 18, dated January 29, XXXX. Refer to Section 11.15 for previous reports prepared for the subject property.

Other Sources

Other sources are those not listed as "standard historical sources" under ASTM E-1527 and include, but not limited to (Refer to Example 4-10):

- Miscellaneous maps;
- Newspaper archives;
- Records in the files and/or personal knowledge of the property owner and/or occupants;
- County atlases and plat maps

Example 4-10:

Based on the documented usage of the subject property as industrial since its construction in 1946 as well as other historical sources dating back to 1927, therefore; county plat maps were not viewed for this property.

SITE RECONNAISSANCE

As discussed in Chapter 2, the custodian and/or the assessment team must perform a site-visit on the subject property.

Name, Date, and Time of Reconnaissance

This section names the custodian and/or assessment team members, and also includes the date, and time of the site reconnaissance (Refer to Example 4-11).

Example 4-11

Mr. Thomas M. Socha, Senior Project Manager, and Mr. John M. Smyth, Staff Engineer conducted the exterior site inspection on July 9, 2000. Qualification statements for the environmental professionals involved in this ESA are included in Section 11.21.

Weather

This section describes the weather conditions at the time of the site-visit (Refer to Example 4–12).

Example 4-12:

On July 9, 2000, the time of reconnaissance, weather conditions were approximately 79° F and partly cloudy. The site was inspected in a meander and search pattern.

Visual and Physical Observations

This section describes the observations made on the subject property and any structures to the extent not obstructed by bodies of water, adjacent buildings or other obstacles (Refer to Example 4-13).

Example 4-13:

The subject property consists of two one-story industrial buildings built in 1946 on approximately 1.18 acres which are serviced by to municipal sewer and water, natural gas, and electricity. There is a rectangular shape asphalt parking lot north of the building. The frontage of the subject property is approximately 186 feet along Anywhere Avenue Street and extends north approximately 277 feet to the fence-line behind the two commercial building. (Refer to Photographic Plates #1 and #2).

The subject property is further described as located in Town one South (T.1S) and Range twelve East (R.12E), Section 22, XXX, XXX County, Michigan.

Current Property Use

This section is used to describe what are the current uses (i.e. industrial, commercial, or residential) of the subject property.

List the processes performed on the subject property. Are the users producing, disposing, or storing hazardous waste on the subject property? (Refer to Example 4-14)

Example 4-14:

Since 1946 two industrial buildings according to City of Somewhere Assessor's records have occupied the subject. The current usage of the subject property consists of paint stripping, and painting of vehicle parts.

Intended Property Use

This section describes the future use intent of the subject property (i.e. industrial, commercial, or residential) once the new owners purchase the property or after the current user is finished renovating the property (Refer to Example 4-15).

Example 4-15:

The intended use of the property is to remain as industrial property.

Utilities

Example 4-16, describes any utility located on the subject property such as:
- Electric
- Heating (i.e. gas, oil, etc.)
- Sanitary sewer
- Water mains

Example 4-16:

Public utilities that service the subject property include electric and phone which are both located overhead. According to the City of Somewhere Water and Sewage Department records, water hook-up for the subject property occurred in 1946 and the sanitary sewer hook-up occurred during the 1920's. Heating for the subject property's buildings is accomplished through natural gas, forced air, and steam.

Water Wells

This section describes any past or current water wells on the subject property (Refer to Example 4-17).

An improperly abandoned well might become used as an inexpensive disposal site for hazardous waste, brine, or human waste.

Example 4-17:

The City of Somewhere Water and Sewage Department records indicate that water hook-up occurred in 1946 for the current subject property building. PT has not identified through review of standard and other historical sources, the historical existence of a private well on the subject property.

Sanitary Sewer/Septic System

This section describes when the sanitary sewer or septic system was connected and installed on the subject property (Refer to Example 4-18).

Like wells abandoned septic tanks might become used as an inexpensive disposal site for hazardous waste, brine, or human waste.

Example 4-18:

A sanitary sewer has connected the building since the 1920's according to the City of Somewhere Water and Sewage Department records.

BUILDING INTERIOR

This section describes the interior of any structures on the subject property. Note the means of heating and cooling of the buildings on the property, including the fuel sources for heating and cooling shall be identified in this section (for example, heating oil, gas, electric, radiators from steam boiler fueled by gas).

Observe any signs of staining or corrosion on floors, walls, or ceilings, and note all drains and sumps in connection with the subject property (Refer to Example 4-19).

Example 4-19:

There are two industrial buildings located on the subject property 1111 Anywhere Avenue and 1112 Anywhere.

1111 Anywhere

The 1946 constructed building located at 1111 Anywhere Avenue and is approximately 7,976 square feet in size and consists of an office area located in the east side of the building and the main shop area. The building is composed of a steel frame on a concrete foundation with concrete block walls. The office area includes a combination of 9" x 9" black vinyl title and carpet covered flooring, the interior walls consists of concrete block, 2' x 4' inlaid and

12" x 12" ceiling tiles. There is a suspect asbestos containing material covering over the steam pipes. There are two bathrooms located in the office area.

During the power wash process wastewater and sludge enters the flow drain in the floor and then enters into the underground cement holding tank. The sludge is stored on-site for a period of one year and then is disposed of off-site. The wastewater is discharge into the sanitary sewer after laboratory analysis (Photographic Plates #11–12). The City of Somewhere Assessor's records indicated that the there was once a boiler located approximately 40 feet from the south side of the building and approximately 80 feet from the east wall.

The one-story 1950 building addition is approximately 4,026 square feet in size and is composed of a steel joists frame on a concrete foundation with concrete block walls, concrete floors, and a flat roof with composite gypsum roof materials. There are two emergency holding tanks located on the western wall. There is one 480-gallon tank with a two-percent solution of HCL. There is one 480-gallon tank with a two-percent of NaOH solution. These solutions in these tanks are utilized in the power-wash process.

The one-story 1952 building addition is approximately 3,880 square feet in size and is composed of a steel joists frame on a concrete foundation with concrete block walls, and a flat roof with composite gypsum roof materials. The northern side of the addition appears to have oil stained floor from a 55-gallon drum. There is a potential for this oil to migrate into the soil and/or groundwater from cracks in the concrete floor. Therefore, this is a recognized environment concern (Refer to Photographic Plate #5). City of Somewhere assessor's records indicated past history indicated that this addition was heated by forced hot air by oil.

The one-story 1957 building addition is approximately 3,454 square feet in size and is composed of a steel joists frame on a concrete foundation with concrete block walls, and a flat roof with composite gypsum roof materials. This addition contains the Chemical Strip Process with two 240-gallon tanks containing a stripping solution of HCL and NaOH (Refer to Photographic Plates #6—#8). There is evidence of rusting of the tanks and spillage and staining on the north wall of this area. Also, the floor in this area appears to contain residue from HCL and NaOH solution. There is the potential for HCL and NaOH to migrate into the soil and groundwater through cracks in the concrete and therefore, represent a recognized environmental concern to the subject property

The one-story 1973 building addition is approximately 2,440 square feet in size and is composed of a steel joists frame on a concrete foundation with concrete block walls, and a flat roof with composite gypsum roof materials. There are two pyrosis bake ovens that are located at this addition (Refer to Photographic Plates #9—#10). The painted parts are bake until the paint is a powder and then are power washed off.

1112 Anywhere Avenue

The 1946 one-story original building is composed of a steel frame on a concrete foundation with concrete block walls, and a flat roof with composite gypsum roof materials. The paint priming of vehicles and paint booth processes are located in this building (Refer to #13 and #15).

The one-story 1951 building addition is composed of a steel frame on a concrete foundation with concrete block walls, and a flat roof with composite gypsum roof materials.

The one-story 1955 addition is composed of a steel frame on a concrete foundation and concrete block walls, and a flat roof with composite gypsum roof materials.

The one-story 1956 addition is composed of a steel frame on a concrete foundation and concrete block walls, and a flat roof with composite gypsum roof materials. This addition contains the sandblasting room where vehicle parts are sandblasted (Refer to Photographic Plates #16).

Interior Waste Streams

This section identifies the wastes streams being produced by the subject property, and where and how the waste is being disposed (Refer to Example 4-20).

Example 4-20:

Solid waste stream is sand from sand blasting and sludge and wastewater from a power wash station. The sand blasting material is recycled until it could no longer be utilized for sand blasting parts and then is disposed in non-hazardous solid waste dumpster and transported off-site by XXX Waste, Inc.

Sludge and waste water is collect in an underground concrete pit and then is discharged into the City of Somewhere sewer after laboratory analysis determines its contaminates level.

BUILDING EXTERIOR

While on the site-visit, view areas of possible avenues of past and current storage, production, disposal of hazardous substances and petroleum products (Refer to Example 4-21).

Example 4-21:

The entrance to the subject property is towards the east from Anywhere Avenue Street. There is an asphalt-paved parking lot that is approximately 3,500 square feet in size located on 1111 Anywhere Avenue Street. The building exteriors located at 1111 and 1112 are comprised of brick, and concrete block building materials.

There is an asphalt alleyway between 1111 and 1112 Anywhere Avenue that is approximately 150 feet in length from Anywhere Avenue Street to the truck well. There is also a vehicle and materials storage area behind 1112 that is approximately 5,625 square feet in size.

Exterior Site Operations

This section describes any exterior industrial or commercial process that is taken place on the subject property (i.e. metal recycling from old automobiles, wastewater treatment, etc.) (Refer to Example 4-22).

Example 4-22:

No exterior site operations exist for the subject property other than a parking lot and a truck well for the industrial building located at 1111 Anywhere Avenue.

Hazardous Waste Disposal

This section describes any history or current use of hazardous waste produced and how it is/was being stored and disposed in connection with the subject property (Refer to Example 4-23).

Example 4-23:

There has been no historic hazardous waste identified in relation to the subject property area through review of reasonably ascertainable standard and other historic sources consulted and review of current uniform hazardous waste manifests, waste water discharge permits, and air mission permits indicates the does not generate hazardous waste.

Agricultural Waste

This section describes any past or current agricultural waste in relation to the subject property (Refer to Example 4-24).

Many agricultural lands often have a waste dump which was used by the farmer and his family. This dump might contain pesticides, herbicides, household hazardous materials, and oil products (e.g. oil filters from farm equipment).

Example 4-24:

There has been no current or historic agricultural waste identified in relation to the subject property area through review of reasonably ascertainable standard and other historic sources consulted.

Solid Waste Stream and Disposal

This section describes what process produces solid waste and how is that waste being stored and disposed in connection with the subject property (Refer to Example 4-25).

Example 4-25:

Disposal of non-hazardous solid waste is collected for off-site disposition by XXX Waste Inc. PT has not identified record of land filling, dumping, and/or fill material of unknown origin or suspect environmental concern on the subject property.

Underground and Above Ground Storage Tanks

This section describes past or current uses of underground (USTs) and/or above ground storage tanks (ASTs) on the subject property (Refer to Example 4-26).

Example 4-26:

Review of reasonably ascertainable standard and other historical sources, and site observations, indicated that there have been no USTs or ASTs on the subject property. The City of Somewhere Assessor's records indicated that the subject property building located at 1111 Anywhere once contained a boiler and the 1952 building addition was heated by forced air—oil. Boilers usually utilizes oil as a fuel source and therefore, an UST could be still present on the subject property and the potential exist for leakage from the UST which can migrate into the soil and groundwater and is a recognized environmental concern. Thus, this is a recognized environmental concern.

Pits, Ponds, Lagoons and Waste Disposal Areas

Pits, ponds, lagoons and waste disposal areas are described in this section if located on the subject and adjoining properties, and especially if they are used in connection with waste disposal or waste treatment on the subject and adjoining properties (Refer to Example 4-27).

Example 4-27:

No pits, ponds, or lagoons were identified as current obstacles.

Polychlorinated Biphenyls

This section describes any electrical or hydraulic equipment (i.e. transformers) that are known to contain polychlorinated biphenyls (PCBs) or likely to contain PCBs on the subject property (Refer to Example 4-28).

The regulatory requirements are contained in federal PCB regulations Toxic Substances Control Act, 40 CFR 761 and any state or local regulations must be kept on file at the facility. Federal PCB regulations generally provide for the continued use of PCB-containing equipment in a closed electrical system for the remainder of its useful life at an industrial facility.

PCB-containing equipment is generally classified into three groups:

- PCB equipment containing greater than 500 parts per million by weight (ppm) of PCB;
- PCB-contaminated equipment containing greater then 50 ppm PCB but less than 500 ppm; and
- Non-PCB materials containing less than 50 ppm PCB. These three groups have different requirements related to their use and disposal. Dilution as a method of changing classifications is strictly prohibited.

Each facility should have on file a drawing showing the location of all PCB transformers, capacitors, and storage areas. The drawing should be updated whenever a change is made. All PCB transformers, PCB transformer locations, PCB containers, large PCB capacitors and PCB storage areas must be marked with a specific PCB label.

A thorough inventory of all PCB-containing equipment and retrofitted PCB equipment at the facility must continue to be maintained. This must include PCB equipment in use, in storage for reuse, or in storage for disposal. These inventories must be updated annually by July 1st for the proceeding year, and must reflect prior years inventories less those PCB items which have been removed from service and/or disposed off-site. Annual inventories from 1978 to the present are required to be on file at the facility.

Retrofitted transformers initially require the same record keeping as PCB transformers. The transformers cannot be reclassified to non-PCB until one full calendar year of analytical data indicating less than 50 ppm of PCBs has been achieved. A non-PCB dielectric must not be put into a transformer until it is reclassified non-PCB.

EPA regulations required that PCB transformers be registered with local fire departments by December 1, 1985, and that transformers in or near commercial buildings also be registered with the building owners by that date. Since December 1, 1985, it has been illegal to store combustible materials within five meters of a PCB transformer enclosure (or of the transformer itself it unenclosed).

Example 4-28:

There has been current or historic oil-cooled electric or hydraulic equipment, or other suspect PCB containing equipment identified in relation to the subject property area through review of reasonably ascertainable standard and other historic sources consulted. The subject property was inspected for oil-cooled electric or hydraulic equipment, or other suspect PCB containing equipment.

The 1995 Phase I ESA had identified two hydraulic hoists that were located in the 1112 Anywhere Avenue building. These two hoists were removed and soil samples were taken and analyzed during the 1995 Phase II. The analytical results indicated the soil samples were below Part 201-Commercial III Residential Drinking Water Protection Criteria.

The during the July 9, 2000 site reconnaissance a two-pole PCB transformer was located on the north side of 1112 Anywhere Avenue both transformers appears be to in good condition with no evidence of leakage, corrosion, or spillage associated with these transformers.

Electromagnetic Fields

This section describes any electromagnetic field located on the subject and adjoining properties (i.e. electric power substations, transformers, etc.) (Refer to Example 4-29).

Certain types of cancers (i.e. leukemia and brain) may be linked to electromagnet fields according to some epidemiological studies.

Example 4-29:

No electromagnetic fields were identified.

Vegetation

This section describes any noticeable stressed or dead vegetation on the subject property (Refer to Example 4-30).

Lack of vegetation in a spot(s) on the subject property may be due to poisoning by herbicide, hazardous materials or materials that hinders plant growth.

Example 4-30:

There was no noticeable stressed or dead vegetation on the subject property.

Oil and Gas Wells

This section describes any oil and gas wells on the subject and adjoining properties (Refer to Example 4-31).

Oil and gas wells may contaminate the groundwater to an undrinkable level due to the increase production of brine from the wells.

Example 4-31:

PT contacted Ms. Joan Doe, Michigan Department of Environmental Quality—Geologic Survey Division (MDEQ-GSD) on July 15, 2000. Ms. Joan Doe reviewed the MDEQ's oil and gas program database regarding any oil and gas wells located in the vicinity of the subject property. No oil and gas wells are located on or adjacent to the subject property or identified in the subject property area through review of reasonably ascertainable standard and other historic sources consulted.

Topography

Topography is use to determine the migration of contaminates from adjoining properties and surrounding area onto the subject property (Refer to Example 4-32).

Example 4-32:

The United States Geological Survey Division (U.S.G.S.) 7.5 Minute Topographic Map (XXX, Michigan Quadrangle, 1968, Photorevised 1983) (Section 11.5) for the subject property was reviewed in accordance with the ASTM standards to evaluate potential migration pathways of substances. The map was reviewed to determine if conditions exist whereby hazardous substances or petroleum products migrate to or from the subject property to surface water, groundwater or soil.

The subject property and immediate surrounding area appears to slope to the east/southeast towards the Somewhere River.

Drainage Patterns

This section describes the direction of surface water runoff and wastewater, other liquid (including storm water), any discharge into a drain, ditches, or steams on or adjacent to the property (Refer to Example 4-33).

Example 4-33:

The property appears to slope east/southeast towards the Somewhere River.

General Soil Profile

This section describes the soil types that are present on the subject property. This is important in determining the migration of any pollutants thru the soil and onto the subject property (Refer to Example 4-34).

Example 4-34:

According to information available from the United States Department of Agriculture, Soil Survey of XXX County, Michigan (published 1977) the subject property is characterized by predominately the following soil type: Pewamo-Blount-Metamora. This soil map unit is identified in the Wayne County Soil Survey as very poorly drained to somewhat poorly drained and nearly level to gently sloping soil, which is subject to frequent flooding. Permeability is moderately rapid in the sandy or loamy upper part of the subsoil and moderately slow in the lower part and the underlying material. The available water capacity is high. Runoff is slow.

Geology

Geology is the science that studies the history of the earth and its life especially as recorded in rocks. Geology is one source in determining the extent of migration of a pollutant to the subject property from the adjoining properties and surrounding area (Refer to Example 4-35).

Example 4-35:

The "Hydrogeologic Atlas of Michigan" produced by Western Michigan University and the United States Environmental Protection Agency, 1981, as well as, "Quaternary Geology of Southern Michigan, Department of Geological Sciences, University of Michigan, Ann Arbor, Michigan (1982)," and "Bedrock Geology of Southern Michigan, Michigan Department of Natural Resources (MDNR), Geological Survey Division (1987)," were reviewed.

According to these sources, in this area of XXX County, quaternary deposits are underlain by Dundee Limestone bedrock geology of Devonian age. Bedrock is overlain by glacial drift material generally characterized as lacustrine clay and silt. The thickness of glacial deposits ranges from one to ten meters.

Local Groundwater Flow

This section describes but not limited to the direction of the groundwater flow and rate(s), the depth of the groundwater encountered below the ground of the subject property, and the quality of the groundwater and is it used for drinking water (Refer to Example 4-36).

Example 4-36:

In absence of site-specific information, a review of topographic features/gradient may be suggestive of the hydrogeologic gradient in the immediate subject property area, since in a general way, the water table typically conforms to surface topography. Groundwater flow direction is expected to be to the south towards the Somewhere River.

Legal Description of Property

Legal descriptions help describe the location and size of the subject property (Refer to Example 4-37).

Example 4-37:

The subject property is located in the City of XXX, XXX County, Michigan. The subject property consists of Lot 64 S100' E179.55' in subdivision Plat of Park Lots. The subject property is also located in Parcel Ward X, Item number 1809. The subject property is approximately 179.55 ft in length and 100 ft. in width located between XXX and XX. The current property at XXX Woodward Avenue serves as XXX Hall and the owner is XXX Hall, Inc.

ADJOINING PROPERTIES

This section describes the adjoining properties uses from record reviews and from the subject property site-visit (Refer to Example 4-38).

Example 4-38:

A visual inspection of the adjoining properties was made from the subject property and public thoroughfares. Additionally adjoining properties were reviewed through other sources consulted throughout this report.

North Adjoining Property

The north adjoining property are railroad tracks located approximately 30 feet from the subject property. Beyond the railroad tracks approximately 100 feet is the XXX Transport Company a trucking company (Photographic Plate #16).

South Adjoining Property

The south adjoining property currently contains the XXX Architecture Materials. This building was built since the 1940's according to Sanborn maps (Photographic Plate #17).

East Adjoining Property

This property currently contains XXX Chrome and Chemical Plant located at 8611 ABC and is listed as a CERCLIS site. The U.S. EPA Region 5 office was FOIA and PT is still waiting for their response. Until further data can confirm otherwise this site represent a recognized environmental concern (Photographic Plate #18).

West Adjoining Property

The west adjoining property currently contains a vacant lot.

Through site observations, regulatory records review, specific inquiry with the MDEQ, Storage Tank Division (STD), Somewhere, Michigan and U.S. EPA Region 5, and through the course of review of reasonably ascertainable standard and other historical sources for the subject property and area, PT has identified that the adjoining properties located at 8611 ABC and 9751 Anywhere Avenue represents a potential off-site concern to the subject property.

REGULATORY FILE REVIEW

Regulatory records can determine whether or not a subject property is listed as a site with recognized environmental conditions. The records also examine the adjoining properties and surrounding area, they also determine if these sites are generators of hazardous waste and have or had leaking underground storage tanks (Refer to Example 4-39 and 4-40).

Regulatory records review can reveal the following:

- The distance from the subject property to RCRA generators and hazardous materials generators from the surrounding area.
- The chemicals (types and quantities) used on the subject property and surrounding area.
- The enforcement actions taken from chemicals that might have been released into the environment.

- Leaking tanks may be close enough to the subject property to be of concern.
- The recognized environmental condition might be upgradient (in terms of groundwater movement) to the subject property.

Example 4-39:

Search at a 1.0 Mile Radius

NPL	Federal National Priorities List
RCRIS—TSD	RCRA Information System TSDF
SHWS	State Hazardous Waste Sites
RCRIS—Generators	RCRA Information System Small and Large Quality Generator

Search at a 0.5 Mile Radius

CERCLIS	Comprehensive Environmental Response, Compensation, and Liability Information System
LUST	State Leaking Underground Storage Tank List
SWF/LF	State Solid Waste Facilities
UST	State Registered Underground Storage Tanks

Search Subject Property

ERNS	Emergency Response Notification System
UST	State Registered Underground Storage Tanks
RCRA—Generators	RCRA Information System Small and Large Quality Generator

Search Adjoining Property

UST	State Registered Underground Storage Tanks
RCRA—Generators	RCRA Information System Small and Large Quality Generator

FEDERAL RECORDS FROM THE UNITED STATES ENVIRONMENTAL PROTECTION AGENCY

- **CERCLIS: Comprehensive Environmental Response, Compensation, and Liability Information System**

CERCLIS contains data on potentially hazardous waste sites that have been reported to the USEPA by states, municipalities, private companies and private persons, pursuant to Section 103 of the Comprehensive Environmental Response, Compensation, and Liability Act (CERCLA). CERCLIS contains sites that are either proposed to or on the National Priorities List (NPL) and sites which are in the screening and assessment phase for possible inclusion on the NPL. The date on the CERCLIS list used was September 3, 1996.

There is no listing found for the subject property and no listings for properties in the 0.5-mile radius search.

- **NPL: National Priority List**

The NPL is a subset of CERCLIS and identifies over 1,200 sites for priority cleanup under the Superfund Program. The date on the NPL list used was July 1 1996.

There is no listing found for the subject property and no listings for properties in the 1.0-mile radius search.

- **ERNS: Emergency Response Notification System**

ERNS records and stores information on reported release of oil and hazardous substances. The date on the ERNS list used was March 31, 1996.

There is no listing for the subject property.

- **RCRIS: Resource Conservation and Recovery Information System**

RCRIS includes selective information on sites which generate, transport, store, treat and/or dispose of hazardous waste as defined by the Resource Conservation and Recovery Act (RCRA). The date on the RCRIS list used was May 31, 1996.

There are no listings for the subject and adjoining properties. There are no listings of RCRA TSD facilities in a 1.0-mile radius. However, there are RCRIS-SQG and RCRIS-LQG within a 0.5-mile radius. These sites are listed below.

Site	Location
XXX Station #53XX	3555 XXX Avenue
City of XXX	433 XXX Street
XXX Hospital	3990 XXX
XXX Public Schools	444 XXX
XXX Telephone Co.	52 XXX
XXX Towers XXX	4 XXX
XXX. XXX Heating	42 XXX

STATE RECORDS FROM THE MICHIGAN DEPARTMENT OF ENVIRONMENTAL QUALITY

- **SHWS: State Hazardous Waste Site**

SHWS records are Michigan's equivalent to CERCLIS. These sites may or may not already be listed on the federal CERCLIS list. Priority sites planned for cleanup using state funds (state equivalent of Superfund) are identified along with sites where potentially responsible parties will pay for cleanup. The Act 307 Site list was dated April 1, 1995 from MDEQ Emergency Environmental Response Division.

There are no listings for the subject and adjoining properties. However, there is one site within the 1.0-mile radius as listed below.

Site	Location
XXX Station or XXXl #05-XXX	6666 XXX Avenue

- **SWF/LF: Michigan Solid Waste Facilities**

SWF/LF type records typically contain an inventory of solid waste disposal facilities or landfills in a particular state. Depending on the state, these may be active or inactive facilities or open dumps that failed to meet RCRA Section 2004 criteria for solid waste landfills or disposal sites. The Michigan Solid Waste Facilities list was dated February 29, 1996 from MDEQ Waste Management Division was reviewed.

There are no solid waste disposal facilities or landfills within the 0.5-mile radius.

- **LUST: Leaking Underground Storage Tank Sites**

LUST records contain an inventory of reported leaking underground storage tank incidents. The date on the Leaking Underground Storage Tank Sites list was April 1, 1995 and the list was obtained from MDEQ UST Division.

There are no LUSTs on the subject property. However, the following sites are sites within the 0.5-mile radius.

Site	Location
XXX's (Former)	25 or 77 XXX Street
XXX Station #5XXX	5555 XXX Avenue
City of XXX	433 XXX Street
XXX Patient Family Services	1111 XXX Avenue
XXX Investment Co.	4444 XXX Avenue
XXX V.A. Hospital	1113 XXX Street
XXX#05-XXX or XXX N Go	6666 XXX Avenue

- **UST: Michigan UST**

UST's are regulated under Subtitle I of the Resource Conservation and Recovery Act (RCRA) and must be registered with the Michigan Department of State Police, Fire Marshall Division. The date of the Michigan UST list was May 1, 1996 from the MDEQ UST Division.

There are no USTs located at the subject property. However, the following sites are within the 0.5-mile radius.

Site	Location
XXX Station #5XXX	5555 XXX Avenue
XXX Columbia Bldg.	XX XXX Street
XXX's Building	77 or 25 XXX Street
XXXHospital	9990 XXX
American XXX	100 XXX Avenue
XXX Car Service	467 XXX Avenue
XXX Armored Inc.	467 XXX
Fire Dept.—Engine #5	434 XXX
XXX Institute, Inc.	222 XXX.
XXX Public Schools	444 XXX

Example 4-40:

Appendix 11.8 includes the VISTA Information System (VISTA) Site Assessment Report, San Diego, California. This report details facilities found in state and federal environmental databases which are located within a pre-specified distance {i.e., Approximate Minimum Search Distance (AMSD)} from the subject property. The pre-specified distances are established per the ASTM Standard (E-1527-00). ASTM has established search distances for federal and state sites of recognized environmental concern and the distances at which particular types of liabilities (e.g., NPL, CERCLIS, RCRA, etc.) are considered to have potential environmental impact to the subject property. Appendix A contains a brief description of each database.

Not all sites found in the VISTA report are located within the AMSD as specified by ASTM. In addition, PT may identify sites identified within the Unmappable Section of the Vista report as within AMSD, by review of additional site location information. Additional sites may also be identified through consultation with State of Michigan Department of Environmental Quality site for registered UST and Part 201 Natural Resources and Environmental Protection Act (NREPA), P.A. 451, 1994, as amended sites. The following is a listing of sites found with the AMSD.

Government Record	AMSD in Miles	Number Found
NPL		
EPA National Priority List (NPL)	1.0	0
CERCLIS		
EPA Comprehensive Environmental Response Compensation and Liability Information System (CERCLIS) List	1.0	3

XXX Chrome Chemical Company, 8611-35 ABC Avenue, Somewhere, Michigan— This site is located less than one-eight mile east of the subject property. According to VISTA, this site is a CERCLIS-NFRAP. A NFRAP is sites may be sites where, following an initial investigation, no contamination was found, contamination was removed quickly or the contamination was not serious enough to require Federal Superfund action or NPL consideration. The VISTA search has this site as a status of not on NPL.

Based on the distance to the subject property this site does appear to represent a recognized environmental concern to the subject property.

XXX/Acustar Somewhere Axle, 6700 Brick Road, Somewhere, Michigan—This site is located between one-half mile and one mile west of the subject property. According to VISTA, the preliminary assessment has no further remedial action planned for this site. Based on distance, regional geology and groundwater flow direction to the south and the site is located side-gradient to the subject property, this site does not appear to represent a recognized environmental concern to the subject property.

Waste Acid and Chemical, 6520 Anywhere, Somewhere, Michigan—This site is located between one-half mile and one mile southwest of the subject property. According to VISTA, interim response is in progress. According to VISTA, the preliminary assessment has no further remedial action planned for this site. Based on regional geology and groundwater flow direction to the south and the site is located down gradient of the subject property, this site does not appear to represent a recognized environmental concern to the subject property.

State Master Lists

State Lists of Hazardous Waste Sites   1.0   0
Identified for Investigation or Remediation-State Equivalent to NPL

State Lists of Hazardous Waste Sites   1.0   0
Identified for Investigation or Remediation-State Equivalent to CERCLIS

TSDFs

Federal Resource Conservation and   1.0   3
Recovery Act (RCRA) Corrective Action Report (CORRACTS) facilities list

XXX Chrome Chemical Company, 8611-35 ABC Avenue, Somewhere, Michigan—This site is located less than one-eight mile east of the subject property. According to VISTA, this site is a low prioritization. Based on distance to the subject property this site does appear to represent a recognized environmental concern to the subject property.

XXCO Inc, 6520 Anywhere, Somewhere, Michigan—This site is located between one-half mile and one mile west of the subject property. According to VISTA, there is a low prioritization status for this site. Based on distance, regional geology and groundwater flow direction to the south and the site is located side-gradient of the subject property, this site does not appear to represent a recognized environmental concern to the subject property.

XXX Chemical, 6451 Anywhere Avenue, Somewhere, Michigan—This site is located between one-half mile and one mile southwest of the subject property. According to VISTA, for this site prioritization status is not reported and no further corrective action is expected at this time. Based on regional geology and groundwater flow direction to the south and the site is located down gradient of the subject property, this site does not appear to represent a recognized environmental concern to the subject property.

Federal Resource Conservation and Recovery Act (RCRA) non-Corrective Action Report (CORRACTS) Treatment, Storage and Disposal Facility (TSD) list   1.0   0

RCRA Generator

EPA Resource Conservation and Recovery Act (RCRA) Notifiers List   0.25   5

XXX Chrome Chemical Company, 8611-35 ABC Avenue, Somewhere, Michigan—This site is located less than one-eighth mile southeast of the subject property. According to VISTA, this is a large quantity generator that generates at least 1,000 kilograms per month of non-acutely hazardous waste (or 1 kilogram per month of acutely hazardous waste). Based on distance this site does appear to represent a recognized environmental concern to the subject property.

XXX Corp XXX Manufacturing Division, 8616 ABC Michigan. This site is located one-eighth mile southeast of the subject property. According to VISTA, this is a small quantity generator that generates at least 100 kilograms per month but less than 1000 kilograms per month of non-acutely hazardous waste. Based on regional geology and down gradient of the subject property, lack of off-site contaminant migration towards the subject property due to groundwater flow direction to the south, this site does not appear to represent a recognized environmental concern to the subject property.

XXX Metals Corp Plant 2, 8250 ABC, Somewhere, Michigan. This site is located one-eighth mile southwest of the subject property. According to VISTA, this is a small quantity generator that generates at least 100 kilograms per month but less than 1000 kilograms per month of non-acutely hazardous waste. Based on regional geology and groundwater flow to the south, and the site down gradient from the subject property this site does not appear to represent a recognized environmental concern to the subject property.

XXX Tool Die Company, 8818 XXX, Somewhere, Michigan. This site is located one-eighth mile southeast of the subject property. According to VISTA, this is a large quantity generator that generates at least 1,000 kilograms per month of non-acutely hazardous waste (or 1 kilogram per month of acutely hazardous waste). Based on regional geology and side-gradient of the subject property groundwater flow direction to the south/southeast, this site does not appear to represent a recognized environmental concern to the subject property.

XXX Electric Company, 8180 Brick Rd, Somewhere, Michigan. This site is located between one-eighth and one-quarter mile north of the subject property. According to VISTA, this is a small quantity generator that generates at least 100 kilograms per month but less than 1000 kilograms per month of non-acutely hazardous waste. Based on regional geology and hydrogeology this site does not appear to represent a recognized environmental concern to the subject property.

ERNS

Federal Emergency Response Notification System (ERNS) List	0.25 Mile	0

LUSTs

Michigan Leaking Registered Underground Storage Tank (LUST) List	0.125	2

XXX Architectural Materials, 9751 Anywhere, Somewhere, Michigan—This site is located less than one-eighth mile southeast of the subject property. According to VISTA, this is an open LUST site. Based on the review of MDEQ file indicates that the concentrations are below Tier I residential soil volatilization to indoor air inhalation risk based screening levels (RBSLs) and groundwater concentrations are below Tier I Residential health-based aesthic criteria RBSLs. Therefore, the file review indicates that the concentration in soil and groundwater are below applicable Tier I Residential Cleanup Criteria and therefore a site closure should be submitted to the MDEQ. This site does not appear to represent a recognized environmental concern to the subject property (Refer to Section 11.14 For MDEQ file).

XXX Electric Company, 8180 Brick Rd, Somewhere, Michigan. This site is located between one-eighth and one-quarter mile north of the subject property. According to VISTA, this is an open LUST site. Based on the review of MDEQ file indicates that the concentrations are below Tier I residential soil volatilization to indoor air inhalation risk based screening levels (RBSLs) and groundwater concentrations are below Tier I Residential health-based aesthic criteria RBSLs. Therefore, the file review indicates that the concentration in soil and groundwater are below applicable Tier I Residential Cleanup Criteria and therefore a site closure should be submitted to the MDEQ. Also, based on regional geology and hydrogeology the groundwater is limited to the saturation zone. Therefore, this site does not appear to represent a recognized environmental concern to the subject property (Refer to Section 11.14 For MDEQ file).

Solid Waste Landfills

Solid Waste Disposal/Landfills (landfill, transfer station) List	1.0 Mile	0

USTs

Michigan Department of Environmental Quality (MDEQ) Registered Underground Storage Tank (UST) List	Subject property & 2 Adjoining

XXX Architectural Materials, 9751 Anywhere, Somewhere, Michigan—This site is located less than one-eighth mile southeast of the subject property. According to VISTA, this site removed one 2,000 gallons gasoline UST. Based on regional geology and hydrogeology this site does not appear to represent a recognized environmental concern to the subject property.

XXX Chrome Chemical Company, 8611-35 ABC Avenue, Somewhere, Michigan— This site is located less than one-eighth mile southeast of the subject property. According to VISTA, this site removed five 3,000 gallons USTs containing hazardous substance, two 6,000 USTs containing, di-is-decylphthalate, and one 6,000 gallons USTs containing hazardous substances.

ODNR

Michigan Department of Environmental Quality Division of Oil and Gas (well logs)	Subject property 0

PT contacted Ms. Joan Doe, Michigan Department of Environmental Quality—Geologic Survey Division (MDEQ-GSD) on July 15, 2000. Ms. Doe reviewed the MDEQ's oil and gas program database regarding any oil and gas wells located in the vicinity of the subject property. No oil and gas wells are located on or adjacent to the subject property or identified in the subject property area through review of reasonably ascertainable standard and other historic sources consulted.

National Heritage Program

Natural Heritage Program (i.e. endangered species and unique habitats)	Subject property & Adjoining	0

The subject property and adjoining properties have been developed land since the 1945 and therefore, this section does not have to be completed.

State Historical Society

State Historical Society for historical/ archeological significance	Subject property & Adjoining	0

The subject property and adjoining properties have been developed land since the 1945 and therefore, this section does not have to be completed.

Local Fire Department   Subject property

The City of Somewhere Fire Department did not respond to PT's written request for information within the time frame this report was prepared. If the Somewhere Fire Department responds in the future and the material changes any conclusion, then the client will be notified immediately.

Local Health Department   Subject property

The City of Somewhere Health Department did not respond to PT's written request for information within the time frame this report was prepared. If the City of Somewhere Health Department responds in the future and the material changes any conclusion, then the client will be notified immediately.

Local Building Department

The local building department has records or permits whenever the user constructs, alter, or demolish improvements on the subject property (Refer to Example 4-41).

Example 4-41:
Subject property

The City of Somewhere Building Department records indicated the same information as found in the City of Somewhere Assessor's records.

Local Zoning Department

The local zoning department has records of current and past permitted uses for the subject property (i.e. is the subject property zoned industrial, commercial, residential) (Refer to Example 4-42).

Example 4-42:
Subject property

Inquiry with the City of Somewhere Assessor's and Building Department Offices, Somewhere, Michigan, has indicated that the subject property is currently zoned as M-2 Light Industrial. The subject property has always had an industrial building on the property since 1946 as indicated by the Sanborn Maps.

Copies of the current and oldest assessment records for the subject property, which were reasonably ascertainable, are included in Section 11.12.

Map

Make a sketch of the subject property and the surrounding area. Note the businesses and industries on the adjoining properties. Label the streets, alleyways, transformers, manholes, drums, ASTs, USTs, etc. (Refer to Example 4-43).

Example 4-43:

A map locating all identified sites and their distance from the subject property is included in Section 11.8.

Chapter 5: From Non-Scope Sources to Documentation of Sources

The topics discussed in this chapter include:
- Non-Scope Sources
- Multi-Media Compliance Inspection
- Conclusions and Recommendations

NON-SCOPE AND ADDITIONAL SOURCES

There may be environmental issues or conditions at a property that parties may wish to assess in connection with commercial real estate that are outside the scope of ASTM E-1527.

The following is a list of possible non-scope considerations:
- Asbestos-Containing Materials;
- Radon;
- Lead-Based Paint;
- Lead in Drinking Water, and;
- Wetlands

Asbestos Containing Materials Inspection

This section describes any sampling or observation of asbestos containing materials (ACM) was performed in connection with the subject property (Refer to Example 5-1).

Asbestos was used in fireproofing of materials that dates backs to the 1890's with roofing materials.

Asbestos products are heat resistant, flexible, and durable, and are commonly found in building construction and insulating materials (e.g., floor tile, fire preventative structures, pipe wrap, etc,). Collectively, asbestos containing products are often referred to as asbestos containing materials or ACM ASTM Standard Designation E-1527-00 defines ACM as containing more than one percent asbestos.

Friable ACM is defined as any material containing more then 1% asbestos, as determined by using the method specified in Appendix A, Subpart F, 40 CFR Part 763, Section 1, Polarized Light Microscopy, that when dry, can be crumbled, pulverized, or reduced to powder by hand pressure.

Non-friable refers to materials that contain asbestos bound by cement, plastic, adhesive, etc., which if handled through routine maintenance, will not become friable.

The subject property is not subject to the Asbestos Hazard Emergency Response Act (AHERA) that relates only to school buildings. *However, the subject property does fall under the National Emission Standard for Hazardous Air Pollutants (NESHAP), which requires that, prior to the commencement of any demolition or substantial renovation operation, the owner or operator must thoroughly inspect the facility/facilities where the operation will occur for the presence of asbestos, including friable and non-friable forms of asbestos.*

All friable regulated asbestos containing materials (RACMs) must be removed from a facility being demolished or renovated before any wrecking or dismantling that would break up the materials. Further, ACM that is determined to be non-friable must be classified as Category I or Category II non-friable asbestos. This classification then determines, based on handling procedures, whether the material must be removed prior to demolition and the preferable means to remove the ACM.

Under the United States Environmental Protection Agency (USEPA) NESHAP regulation, ACMs or any material found to contain asbestos in concentrations greater than one percent, as determined by polarized light microscopy (PLM). RACMs are:

- Friable asbestos material
- Category I non-friable ACM that has become friable
- Category I non-friable ACM that will be or has been subject to grinding, cutting, abmrading
- Category II non-friable ACM that has a high probability of becoming crumbled, pulverized, or reduced to powder by the forces expected to act on the material in the course of demolition or renovation operations regulated by NESHAP.

Category I non-friable ACMs are asbestos-containing packing, gaskets, resilient floor covering, and asphalt roofing products containing more than one percent asbestos.

Category II non-friable materials, excluding Category I non-friable ACM, contain more than one percent asbestos.

Example 5-1:

Any inspection or survey is limited, based on time and cost constraints of the Phase I ESA, to a visual examination of a structure or material to identify materials for sampling and analysis to determine the presence of ACM. Access in building structural columns or equipment liners was not included as part of this limited inspection (e.g., building columns, supports, behind walls, boiler or tank interiors, etc.).

Identification of Certified Asbestos Inspector

PT's accredited asbestos building inspector Mr. Thomas M. Socha did observe suspect friable ACMs during the July 9, 2000 site reconnaissance of the subject property. **Sampling of suspect ACM was completed on July 23, 2000.** TMC Laboratories, Somewhere, Arizona analyzed the suspect ACM bulk samples.

Suspect Non-Friable ACM in Good Condition

The entranceway to 1111 Anywhere Avenue Street contains 9" x 9" black vinyl title that is approximately 100 square feet in size.

Suspect Non-Friable ACM in Damaged Condition

Not applicable.

Analytical Results of Suspect Non-Friable ACM

The 9" x 9" asbestos black floor tiles contain asbestos, however, the black mastic does not contain asbestos materials.

Suspect Friable ACM in Good Condition

The 2' x 4' wormy and 12" x 12" square inch dot ceiling tiles are suspect friable ACM. Pipe insulation covering for the building located at 1111 Anywhere Avenue is also suspect friable ACM.

Analytical Results of Suspect Friable ACM

The pipe insulation has asbestos containing material that is approximately 400 liner feet located in 1111 Anywhere.

Suspect Friable ACM in Damaged Condition

Not applicable.

Analytical Results of Suspect Friable ACM (Damaged

Not applicable.

LEAD BASE PAINT SCREEN

Example 5-2:

A lead base paint inspection was not required for this site refer to Section 11.17.

LEAD IN WATER TESTING

Example 5-3:

A lead inspection in water was not required for this site refer to Section 11.19.

RADON INSPECTION

Example 5-4:

A radon inspection was not required for this site refer to Section 11.18.

WETLANDS INVESTIGATION

Example 5-5:

PT did not observe any obvious wetland plant species or wet areas on the subject property during the July 9, 2000 site reconnaissance.

MULTI-MEDIA COMPLIANCE INSPECTION

Some lending institutions and buyers may want a multi-media compliance inspection performed on the subject property (Refer Example 5-6).

A multi-media compliance inspection includes, but not limited to:

- PCB Management;
- Wastewater Discharge;
- Air Emissions;
- Oil Spill Control;
- Hazardous Waste Generation;
- Community Right-to-Know;
- Hazardous Communication;
- Employee Safety;
- Pesticide Management;
- Underground Storage Tanks;
- Hazardous Materials Storage;

- Drinking Water, and
- Solid Waste Disposal

Example 5-6:

The subject property is not a large quantity generator (LQG) or a small quantity generator (SQG) of hazardous waste; however, the lending institution requested a multi-media compliance inspection.

Air Emissions

Review of XXX County Department of Environment—Air Quality Management Division (XCDOE) records revealed the following information concerning the subject property at 1111 Anywhere. There were no records regarding 1112 Anywhere.

In February 1996, XCDOE issued a certificate of operation to install a pyrosis burn-off furnace. In September 1997, XCDOE informed the subject property on its pyrosis oven permit was canceled because it is exempt from installation permit requirements based on Michigan Public Act 451 of 1994, Part 55, Air Pollution Control, Administrative Rule 285 (f). However, permit to install exemption does not preclude the subject property from complying with other applicable administrative rules under the above Act as well as other local, state, and federal regulations.

Waste Water Discharge

The subject property has wastewater discharge permit number 003-063 with the City of Somewhere Water and Sewerage Department. The effective date of the permit is from October 6, 1996 with the expiration date of September 1, 2000. The permit states that samples of the wastewater must be taken every six months. Grab samples are taken from a two-inch PVC pipe protruding from a holding tank located 39 feet north of south wall and 100 feet west of east wall and analyzed by an independent laboratory. Analytical results can be found in Table 2—Liquid Waste Water. Batch discharge information: frequency—two to three batches per week average, four hours per batch, and 3,000 gallons per batch maximum volume.

RCRA Waste Generate and Disposal

According to shipping manifests the subject property generates non-hazardous sludge and wastewater from the painted parts stripping process which includes the following process steps: 1) The painted part is placed into burn-off furnace until the paint is a powder, 2) The part is taken to the power-wash station and rinsed using a solution of two-percent

HCL and two-percent NaOH. When HCL and NaOH reacts it forms a sodium chloride (salt) precipitate and water both are non-hazardous by-products, 3) The waste water and sludge is collected into concrete underground holding pit, 4) Grab samples are taken and then analyzed at an independent laboratory, 5) If analytical data is below the City of Somewhere regulatory limits then the waste water is discharge into the sanitary sewer, 6) If analytical data is above the regulatory limits then corrective actions are implemented, and 7) Sludge is kept in the approximately two-1,000 gallon emergency tanks for one year and then is pumped into a cargo tank truck and transported off-site by XXX Pumping Service to XXX Reduction Systems, Inc. a disposal facility located in Somewhere, Michigan. This waste is classified as an Other Waste with Waste Number 296 a non-hazardous waste on the shipping manifest. This waste was analyzed for contaminates listed in Table 3—Sludge Waste and compared to State of Michigan Regulatory Limits.

The 1950 building addition has the drums storage area for one 55-gallon of HCL, one 55-gallon of NaOH, and one 55-gallon of hydraulic oil. This storage area lacks the proper spill prevention, control, and countermeasures and drum management. Spills were observed around both the HCL and NaOH 55-gallons drums there is the potential to erode the concrete floor and cause contaminants to migrate into the soil and groundwater. Therefore, this represents a recognized environmental concern to the subject property (Photographic Plate #19—#21).

CONCLUSIONS AND RECOMMENDATIONS

The ASTM standard states that the findings and conclusions section must have the following (Refer Example 5-7):

"We have performed a Phase I Environmental Site Assessment in conformance with the scope and limitations of ASTM Practice E-1527 of [insert address or legal description], the property. Any exceptions to, or deletions from this practice are described in Section [ ] of this report. This assessment has revealed no evidence of recognized environmental conditions in connection with the property,"
OR,
We have performed a Phase I Environmental Site Assessment in conformance with the scope and limitations of ASTM Practice E-1527 of [insert address or legal description], the property. Any exceptions to, or

deletions from this practice are described in Section [ ] of this report. This assessment has revealed no evidence of recognized environmental conditions in connection with the property except for the following;"

Example 5-7:

Conclusion of the Phase I ESA

PT has completed the Phase I ESA, in general conformance with the scope and limitations of ASTM Practice E-1527-00, of an industrial building located at 1111 Anywhere Avenue Street, Somewhere, XXX County, Michigan.

The multi-media compliance inspection concluded that the subject property appears to be in compliance with all federal, state, and local regulations based on regulatory records review.

This assessment has revealed evidence of recognized environmental conditions associated with the subject property as follows:

- The original building section (1946 construction) located at 1111 Anywhere Avenue currently contains approximately two 1,000 gallons emergency open aboveground steel holding tanks which contain sludge and waste water which is generated at the power wash station. The sludge and wastewater contents have the potential to contain (lead, chromium, and other metals). The sludge and wastewater inside these tanks are held on-site for a period of up to one year and the contents are pumped into a cargo tank truck for off-site transport. Due to the long-term operations of the subject property as a paint stripping facility the potential exists for sludge contaminates to spill, leak, and overflow into cracks in the concrete floor and migrate into the soil and groundwater. Thus, these tanks represent a recognized environmental concern to the subject property.

- The north wall of the original building section (1946 construction) located at 1111 Anywhere once contained a former paint booth. Historic records review indicated that the paint booth was operational from at least 1981 to 1995. Due to past long term operations of this paint booth there is the potential for releases of solvents and other contaminants contained in the paints such as chromium, lead, and other metals to migrate into the soil and groundwater through cracks in the concrete floor and represents a recognized environmental concern.

- The south adjoining property historically has been a lithographic and printing facility from at least 1949 to 1968. Printing and lithographic facilities utilize printing fluids, inks, and solvents in there printing process. Past hazardous waste management practices for these wastes might have involved disposing of the waste on-site (i.e. dumping the waste out a side door onto the dirt ground) and this waste has the potential to migrate into the soil and groundwater of the subject property. Thus, this represents a recognized environmental concern.

- The east adjoining property is listed on the U.S. EPA CERCLIS list according to the VISTA database search. There is the potential for off-site migration of contaminates to the subject property. PT has FOIA the US EPA Region 5 District Office regarding the file on the site until PT receives further information regarding this site it represents a recognized environmental concern to the subject property.

- Two chemical stripping tanks are located against the northern wall in the 1957 building addition of 1111 Anywhere Avenue. Each tank contains approximately 240 gallons of HCL and NaOH solution. It appears that residue from HCL and NaOH covers the concrete floor and there is also evidence that both tanks are stained and rusting. The HCL and NaOH are both corrosive chemicals that have the potential to erode the concrete floor thus, allowing HCL and NaOH to migrate into the soil and groundwater. Therefore, this area represents a recognized environmental concern.

- The storage area behind the 1112 Anywhere Avenue historically and currently has been utilized as a storage area for metal parts, metal shavings, and metal 55-gallon drums. Metal shavings are a by-product of metal fabricators facilities that usually contain cutting oils and other solvents. The metal 55-gallon drums may have contained petroleum products which had the potential to leak, and spill unto the dirt ground and the severely damage asphalt pavement. Thus, these contaminates have the potential to migrate into the soil and groundwater which represents a recognized environmental concern.

- The 1111 Anywhere Avenue building according to historical records once contained a metal cutting and forming operation, and a glass working plant. Metal fabricators and glass works operations usually use cutting oils, solvents, cleaners, resins, other petroleum products, and metals in their processes. These contaminates have the potential to migrate into the soil and groundwater through cracks in the concrete floor and represents a recognized environmental concern.

- There appears to be a lack of proper spill prevention practices located in the 1950 building addition of the 1111 Anywhere Avenue building. This area contains one 55-gallon drum of HCL and one 55-gallon drum of NaOH. There is evidence of recent spillage from both drums onto the concrete floor. The potential exists for HCL and NaOH to migrate into the soil and groundwater through cracks in the concrete floor and represents a recognized environmental concern.

- The 1112 Anywhere Avenue building has a stained floor that appears to be from a spill of an unknown substance. The floor slopes toward the sewer drain located in the southwestern bay. Contaminates have the potential to migrate into the soils and groundwater from cracks in the concrete which represents a recognized environmental concern.

- The City of Somewhere Assessor's records indicated that the 1111 Anywhere Avenue building once had a boiler there is no records to indicate if a tank exists or have been removed from the subject property. There also appears to be an indication that the 1952 building addition was heated by forced air-oil. Boilers usually utilize oil, as a fuel source and therefore, an UST could still be present on the subject property. The UST has the potential to leak and their contents would have the potential to migrate into the soil and groundwater. Thus, this represents a recognized environmental concern.

- The 1112 Anywhere Avenue building has been associated with the Smith Hardware Manufacturing Company in 1954 and as a vehicle maintenance garage according to the City of Somewhere Assessing records. The hardware manufacturers utilize metals, cutting oils and other solvents in the producing their products. Vehicle maintenance garages have been associated with petroleum products and solvents. The potential exists for these contaminates to migrate into soil and groundwater from cracks in the concrete floor caused by the lack of proper spill control and management of these materials and thus, represents a recognized environmental concern.

10.2 Recommendations of the Phase I ESA

These recognized environmental conditions are brought to the attention of the client within the requirements of the ASTM Standard Practice E-1527-00. Verification of the presence or absence of contaminants potentially associated with these recognized environmental conditions may be determined through a Phase II ESA investigation at the request of the client. A Cost/risk decision associated with further investigation of these conditions is the decision of the client.

Documentation of Sources

The ASTM E-1527 practice states, "the report shall document each source that was used, even if the source revealed no findings. Sources shall be sufficiently documented, including name, date request for information was filled, date information provided was lasted updated by source, date information was last updated by original source (if provided other than by original source) so as to facilitate reconstruction of the research at a later date." Example 5-8, lists the documentation used and researched in the report.

Example 5-8:
- On-site reconnaissance/walk-through of the subject property conducted by PT personnel July 9, 2000;
- Interviews (or attempts to interview through written or verbal correspondence) with Ms. Mary Smith of Smith Industrial Services, Somewhere, Michigan, subject property A key site manager; City of Somewhere Assessor's Office, Somewhere, Michigan; City of Somewhere Water and Sewage Department; City of Somewhere Building Department, Somewhere, Michigan; City of Somewhere Health Department, Somewhere, Michigan; Michigan Department of Environmental Quality—Geological Survey Division, Somewhere, Michigan; Ms. Jones, MDEQ-STD, Somewhere, Michigan; Ms. XXX, MDEQ-ERD, Somewhere, Michigan;
- Review of reasonably ascertainable records from City of Somewhere Assessment Office, Somewhere, Michigan, for the subject property;
- Research of reasonably ascertainable R.L. Polk's Directories and Bresser's Cross-Index City Directories for the subject property area; Bresser's Cross-Index City Directories were available from 1976 to 1998; R.L. Polk's Directories were available from 1927 to 1968 for the subject property; the only available coverage for the subject property; Bresser's Cross-Index City Directories, XXX County, Michigan, publication agency is the original source;
- Research and review of reasonably ascertainable aerial photographs from 1966 (scale: 1":1,000'), 1975 (scale: 1":1,000'), 1980 (scale: 1":1,000') and 1995 (scale: 1":1,000'); original source of the aerial photographs is unknown; photographs obtained through the XXX (XXX), Somewhere, Michigan, as a secondary source;
- Site photographs taken at the time of the on-site inspection and visual inspection of adjoining properties from the subject property;
- Review of legal description provided by the City of Somewhere Assessor's Office;
- Review of United States Geological Survey (USGS), 7.5 minute quadrangle map for XXX, Michigan Quadrangle, original issued 1963 and photorevised 1980 (original from U.S. Geological Survey, Denver, Colorado, Michigan Department of Natural Resources (MDNR), Geologic Survey Division, Somewhere,

Michigan, and/or Michigan United Conservation Club (MUCC), Somewhere, Michigan, and available from the archive of PT);

- Review of the "Hydrogeologic Atlas of Michigan" produced by Western Michigan University and the United States Environmental Protection Agency, 1981, "Quaternary Geology of Southern Michigan, Department of Geological Sciences, University of Michigan, Ann Arbor, Michigan (1982)," and "Bedrock Geology of Southern Michigan, Michigan Department of Natural Resources (MDNR), Geological Survey Division (1987)," (original sources and available from archive of PT);

- Review of U.S. Department of Agriculture, Soil Survey of XXX County, Michigan, March 1977 (available in the archives of PT; original source XXX County Soil Conservation Service, XXX County, Michigan);

- Review of federal and state regulatory records as part of the ASTM Standard Environmental Record Sources, provided by VISTA, San Diego, California as a subcontracted database service. The last date of update by VISTA for each respective database is included in the VISTA report;

- Review of City of Somewhere Fire Department, Somewhere, Michigan, file information pertaining to the current or historical presence of USTs, ASTs and fires on the subject property;

- Review of Michigan Department of Environmental Quality—Geologic Survey Division, Somewhere Michigan, file information pertaining to the current or historical presence of oil production wells on the subject property. No records were identified;

- Review of City of Somewhere Building Department, Somewhere, Michigan, construction plans and permit records for the subject property property;

- Request for City of Somewhere Health Department, Environmental Health Division, septic and well log information, environmental inspection and correspondence file information for the subject property; no well log or septic information or environmental inspection and correspondence file information was reasonably ascertainable for review;

- Request for Michigan Department of Environmental Quality—Storage Tank Division, Somewhere, Michigan, file information indicating confirmed or suspected release reports, notice of removal, UST registrations, and inspection and site assessment reports Inspection reports; the MDEQ-STD had no records for the files requested;

Appendix A: Chain of Custody

Customer Chain Of Custody			
Project Manager:		Date:	
Client Name:		Financial Institution Name and Contact:	
Client Address:		Financial Institution Address:	
Client Phone Number:		Financial Institution Phone:	

Relinquished By:	Received By:	Date:	Time:

FILE INVENTORY

APPENDIX B: ENVIRONMENTAL ASSESSMENT CHECKLIST

Phase I ESA Checklist				Date:	Page 1 of
Project Name:			Project Address:		
Project Manager:					
Staff Member:					
Client Name:			Financial Institution Name and Contact:		
Client Address:			Financial Institution Name:		
Client Phone Number:			Financial Institution Phone Number:		
Client Fax Number:			Financial Institution Fax Number:		
Project (FOIA) Contact Checklist	Date Inquire was Sent	Date Material was Rec'd	Address		Contact Person and Phone Number
☐ Waste Management Division					
☐ UST Division					
☐ Emergency Response Division					
☐ MDEQ Oil and Gas Division					
☐ National Heritage Program					
☐ Michigan Leaking Underground Storage Tank Division (LUST)					
☐ EPA Region 5					

				Date:	Page 2 of
Project (FOIA) Contact Checklist	Date Inquire was Sent	Date Material was Rec'd	Address		Contact Person and Phone Number
CITY AND LOCAL OFFICES					
☐ Assessor's					
☐ Building Department					
☐ Engineering Department					
☐ Fire Department					
☐ Aerial Photographs					
☐ Planning Department or Mapping Department					
☐ Health Department					

		Date:	Page 3 of
	LIBRARY RESEARCH		

Topographical Maps:	Report Figure:	Caption For Figure:	
	Quadrangle:	Year:	County:
	Township:	Range:	Section Number:

City Directories:

County Plat Maps:

Aerial Photographs:

Sanborn Insurance Maps:

APPENDIX C: REFERENCES

ASTM
100 Barr Harbor Drive
West Conshohocken, Pennsylvania, USA 19428-2959
Phone: (610) 832-9585
Fax: (610) 832-9555
Internet: http://www.astm.org

Environmental Assessment Association
1224 North Nokomis NE
Alexandria, Minnesota 56308 USA
Phone: (320) 763-4320
Fax: (320) 763-9290
E-mail: *eaa@iami.org*
Internet: *http://iami.org/eaa.html*

Environmental Data Resources, Inc.
3530 Post Road
Southport, CT 06490
Phone: 1-800-352-0050
Fax: 1-800-231-6802

Softshel, Inc.
P.O. Box 10813
Midwest City Okalahoma 73140
Phone: (405) 733-2002
Internet: http://www.envirodisk.com

VISTA
5060 Shoreham Drive
San Diego, CA 92122
Phone: 1-800-767-0403
Fax: (858) 450-6195
Email: info@vistainfo.com
Visit our Web site at http://www.vistainfo.com

BIBLIOGRAPHY

ASTM Designation: E1527-00, *Standard Practice for Environmental Site Assessments: Phase I Environmental Site Assessment Process*

ASTM Designation: E1528-00, *Standard Practice for Environmental Site Assessments: Transaction Screen Process*

Briggs, Church E. *Environmental Site Assessment from the Perspective of a Purchaser of Real Property. http://www.enviroglobe.com.*

Environmental Assessment Association. *Certification and Statement of Limiting Conditions.* Minnesota. 1996

Carter, Michael C. and Zawitoski, Gina M. *Developing a Standard Practice for Phase II Environmental Site Assessments.* ASTM Standardization News. April 1994: 34-39.

Colten, Craig E. *From the Environment.* Real Estate Law Journal. Vol. 22 (1994): 346-354.

Conner, Steve R. *Due Diligence in Property Transfers.* Petroleum Engineer International, February 1992:51-54

Conner, Steve R. *Environmental Issues and the Oil and Gas Operator.* Petroleum Engineer International, March 1992:77-78

Conner, Steve R. *Tracking Environmental Liability in Property.* Petroleum Engineer International, April 1992: 62-63

Crealse, Charles and Hedley, James. *ASTM Guidance: Site Assessments for Commercial Real Estate Transactions.* Professional Safety (1994): 40-41.

Gibby, Daniel J. and Moon, Ralph E. *Environmental Consulting Liability An Uncertain Legal Web.* The National Environmental Journal November/December 1994.

Halloran, Mark. "*Environmental Site Assessments.*" The Practical Lawyer Vol. 41: 61-68.

Hess, Kathleen. *Environmental Site Assessment Phase I: A Basic Guide.* Florida: CRC Press, Inc., 1993.

Howard, Bruce P. *No Easy Answers on Due Diligence.* National Law Journal, November 22, 1993: 26-27.

Jones, Richard D. and Herandez, Myriam E. *Standards for Environmental Assessments.* ASTM Standardization News, April 1994: 30-33.

Kertesz, Louise. *Addressing Environmental Issues.* Business Insurance, May 2, 1994: 37.

Pierro, William J. *When Environmental Due Diligence Enters Phase II.* ABA Banking Journal September 1992.

Softshel, Inc. *Environmental Site Assessment.* Vers. 4.0. Computer Software. Softshel, Inc., 2000.

Spearot, Rebecca Ph.D. *Environmental Decisions: Going Beyond Phase I Assessments.* Site Selection & Industrial Development 39: 6(352)-(357) April 11, 1994.

U.S. EPA. *Asbestos/NESHAP Regulated Asbestos Containing Guidance.* Office of Toxic Substances, Washington, DC, EPA 340/1-90-018

INDEX

Aerial photographs, ix, 28, 41, 43, 52-53, 57, 101

Asbestos-containing materials, 91

AHERA, xiii, 92

Assessment, i, iii, iv, vii, viii, x, xv, 4-5, 7, 15-17, 22-24, 29-31, 33-35, 37, 41, 44, 46-49, 58, 60, 62-63, 80, 83-84, 89, 97-98, 101-102, 105, 109, 111-112

ASTM, xiii, xv, 4-6, 17, 28-32, 35-38, 44-47, 51, 60, 62, 75, 83, 91-92, 97-98, 100, 102, 109, 111-112

CERCLA, xiv, xv, 4, 13, 47, 80

CERCLIS, xiii, 22, 25, 37, 39, 41, 45, 61, 78-81, 83-84, 99

CCC, 6-7

Conclusions, viii, x, 8, 29-31, 34, 38, 43, 46-49, 91, 97

Custodian, 7, 15-17, 23-25, 29, 31, 38, 44, 62-63

Documentation, viii, x, 24, 26, 31, 43-44, 46, 51, 59, 91, 100-101

ERNS, xiii, 25, 41, 45, 79-80, 86

Environmental professional, 5, 7, 31, 44-46

Findings, viii, 1, 8, 16, 18, 24, 29-31, 34, 37-38, 97, 100

Fire insurance maps, ix, 23, 28, 41, 43, 54-55

FOIA, xiii, 16, 39, 44, 78, 99

Geology, x, 28, 42, 76, 84-87, 102

Innocent landowner defense, 4

Interviews, vii, viii, ix, 5-6, 16, 19, 28, 31, 41-42, 58-59, 101

Lead base paint screen, 43, 94

Lead in water testing, 43, 94

Manager, vii, 1, 2, 7, 15, 17, 59, 61, 63, 101, 115

NESHAPS, xiii

NPL, xiii, 41, 79-80, 83-84

Observations, exterior, 18

Observations, interior, 18
Phase I environmental site assessment, i, iii, iv, vii, viii, 4, 16, 29-31, 33-35, 44, 97, 111-112
Questionnaire, 5, 16-17, 19
Radon inspection, 43, 95
RCRIS, xiii, 79-80
Reasonably ascertainable, 5, 17, 29, 36-38, 51-53, 56, 60, 69-71, 73-74, 78, 88-89, 101-102
Recognized environmental condition, 6, 79
Records review, vii, ix, 5, 23, 25, 34, 37, 39, 50, 78, 98
RCRA, xiii, xiv, 11, 22, 25, 41, 43, 78-85, 96
Sanborn Map, 55
Site reconnaissance, vii, viii, ix, 5-6, 16, 31, 37-38, 45, 50, 53, 59, 62-63, 73, 93, 95
Site-visit, vii, 2, 5, 15-19, 23-24, 61-63, 68, 77
Soil, x, 8, 18, 20, 27-28, 30, 34, 38-43, 48, 61-62, 67, 71, 73, 75, 86-87, 97-100, 102
Superfund, xiv, xv, 13, 80-81, 83
SARA, xiv, xv, 26
Title records, 17, 28
Topographic map, 6, 28, 74
TSDFs, 41, 84
USTs, 41, 61, 70-71, 82, 87, 89-90, 102
Water wells, ix, 42, 65
Wetlands investigation, 43, 95

ABOUT THE AUTHOR

Thomas M. Socha is the Senior Project Manager at People Technology, Inc., located in Rochester Hills, Michigan. Mr. Socha has a Master of Science in Hazardous Waste Management from Wayne State University, Detroit, Michigan.

Made in the USA
Lexington, KY
23 January 2014